Modern Algebra

Part 2: Homomorphisms and Quotient Groups

by

Carroll W. Boswell

copyright 2018 by Carroll W. Boswell

ISBN-13: 978-1718793033

ISBN-10: 1718793030

Preface to Modern Algebra part 2

This part of the Modern Algebra sequence follows immediately from the lessons in part 1. It is written in the same style and with the same philosophy as part 1. The numbering of the theorems and definitions continue from part 1 rather than beginning again and so that it could be used seamlessly with part 1 as a single semester introduction to modern algebra at the college level, though it contains too much material to be covered in detail in a single semester. It is assumed throughout that the reader has either read part 1 or else has become informed of that material from another source. Anyone who has not learned that material, or does not remember it, will be at a disadvantage here.

This book naturally splits into four sections, each with its own focus. The first section runs to lesson 10 and introduces isomorphisms, automorphisms, and direct products. Direct products, besides being indispensable material in beginning group theory, provide a large number of examples of isomorphic groups.

The second section of the book begins with the introduction of cosets and proceeds immediately to normal subgroups and inner automorphisms. This material is covered from lesson 11 to lesson 19.

From lesson 20 to lesson 25 the focus is on quotient groups, and the important cluster of theorems that detail how quotients work, particularly the three Isomorphism Theorems.

Finally, from lesson 26 to the end some of the most important subgroups of any given group are introduced: the center, the socle, the derived subgroup, and the Frattini subgroup. The peculiar properties and applications of each of these special subgroups is examined.

In comparison to part 1, part 2 is somewhat more abstract and more concerned with groups in general than with specific groups. Part 1 was intended as a means of fixing the concept of the group clearly in the mind by giving a wide variety of specific examples, as well as introducing some of the common algebraic methods. It was written in order to develop an intuition for groups as well as provide the basic facts. Part 2 builds on the intuition developed in Part 1 to derive some of the more complex and general relationships between groups. Part 2 dwells more on proving theorems than discussing examples, though examples still play their critical role in further developing the intuition that is necessary for true understanding.

Table of Contents

Lesson 1: Isomorphisms	1
Lesson 2: The Direct Product	4
Lesson 3: Direct Products of Cyclic Groups	6
Lesson 4: More Complicated Direct Products	9
Lesson 5: Three Theorems about Direct Products	12
Lesson 6: U_n and the Chinese Remainder Theorem	14
Lesson 7: Automorphisms	16
Lesson 8: $\mathbf{Aut}(\mathbf{Z_n})$	18
Lesson 9: More Complicated Automorphism Groups	20
Lesson 10: Characteristic Subgroups	24
Lesson 11: Cosets	27
Lesson 12: Lagrange's Theorem	29
Lesson 13: Normal Subgroups	32
Lesson 14: Conjugations and Inner Automorphisms	35
Lesson 15: Further Properties of Normal Subgroups	38
Lesson 16: Conjugation in S_n	40
Lesson 17: Group Extensions	42
Lesson 18: The Extremes: Simple Groups and Hamiltonian Groups	44
Lesson 19: The Extremes: Elementary Groups	47
Lesson 20: Quotient Groups	49
Lesson 21: Homomorphisms	52
Lesson 22: The First Isomorphism Theorem	55
Lesson 23: The First and Second Correspondence Theorems	57
Lesson 24: The Second and Third Isomorphism Theorems	59
Lesson 25: The Zassenhaus Theorem	61
Lesson 26: Normalizers	63
Lesson 27: The Center of a Group	65
Lesson 28: Centralizers and the Conjugacy Class Equation	68
Lesson 29: p-Groups	70
Lesson 30: Commutators	72
Lesson 31: The Derived Subgroup	74
Lesson 32: The Frattini Subgroup and Non-generators	77
Lesson 33: The Frattini Subgroup and p-Groups	79
Table of Groups	81
Index of Terms	83
Index of Symbols	87

Lesson 1: Isomorphisms

In learning any new mathematical system of thinking, there are two foundational steps to take at the beginning. First, we must establish the particular family of sets we will be discussing by delineating the properties that define them. Second, we must establish the nature of the functions that map one of these special sets to another of them. In Part 1 of Modern Algebra the first step was taken. There we specified the defining structure of the sets we call groups and examined a large number of examples in some detail. Now in Part 2 we will examine the character of various functions that map one group to another group. As usual in mathematics, knowledge of the functions will give us additional insight into the structure of the sets involved. The functions that map one group to another group are called homomorphisms in general, and as we understand these homomorphisms we will gain unexpected information about the structure of the groups involved.

But we will start, not with homomorphisms in general, but with the simplest homomorphisms. As a minimum, then, what do we desire from the functions that map one group to another? There must be some minimal standards because the range of possible functions from one set to another is simply too vast and includes too many functions that have little or no relationship to the algebra that is built into the sets. Naturally, we will primarily be interested in the functions that are closely connected to that algebra. This is the motivation then for our first definition. In giving definitions and theorems in this part of Modern Algebra, I have elected to continue the numbering of them consecutively from the first part. For the exercises I have reset the numbering to the beginning.

Definition 30: Let $< G, + >$ and $< H, \cdot >$ be two groups. A bijection $\phi: G \to H$ is called an **isomorphism** when it satisfies the property $\phi(g_1+g_2) = \phi(g_1) \cdot \phi(g_2)$. When such an isomorphism exists between the groups G and H we say G **is isomorphic to** H and we write it as $G \simeq H$.

There are several things to notice immediately about this definition. First, since ϕ is a bijection, the order of G must equal the order of H. The condition imposed on ϕ amounts to requiring that ϕ preserves the algebra in G, that H is an algebraic identical twin to G. This is the proper way of discussing equality of two groups. We will no longer talk about two groups being "equal". When two groups are the same as each other, when they are indistinguishable except for the names of the elements and the symbol used to denote the operation, we will say the two groups are isomorphic.

The condition imposed on the bijection ϕ, that $\phi(g_1+g_2) = \phi(g_1) \cdot \phi(g_2)$, can be restated in words this way: we may add two elements in G and then map the result over to H, or we may map the two elements of G over to H and then multiply them in H and the results will be the same using either sequence. I have indicated the operations in G and in H by two unmistakably different symbols so there will be no confusion. In future the operations in G and H will be denoted by the same symbol.

We will now collect together the basic properties of isomorphisms into a single theorem. They are all easy to prove. I will continue to use additive notation for the group G and multiplicative notation for the group H to accentuate their difference.

Theorem 37: Let G and H be groups and $\phi: G \to H$ be an isomorphism. Then the following are true:
   a) if 0 is the identity of G and 1 is the identity of H, then $\phi(0) = 1$.
   b) for any $g \in G$, $\phi(-g) = \phi(g)^{-1}$.
   c) for any $g \in G$, $\phi(ng) = \phi(g)^n$.
   d) if g has order n in G, then $\phi(g)$ has order n in H.
   e) if $K \leq G$, then $\phi(K) \leq H$.
   f) if $\phi: G \to H$ and $\psi: H \to K$ are isomorphisms, then the composition $\psi \circ \phi: G \to K$ is an isomorphism.
   g) If $\phi: G \to H$ is an isomorphism then $\phi^{-1}: H \to G$ is also an isomorphism.

Proof: a) $\phi(g) = \phi(g+0) = \phi(g)\cdot\phi(0)$ for any $g \in G$. Therefore $\phi(0)$ must be the identity element in H.

b) We simply compute the product of $\phi(g)$ and $\phi(-g)$ and observe that we get the identity element of H: $\phi(g)\cdot\phi(-g) = \phi(g + -g) = \phi(0) = 1$ by part a.

The proofs of parts c through g are left as exercises.

Theorem 37 is exactly what we would expect from a function that is preserving algebraic properties. Isomorphisms preserve the subgroups and the interrelationships between the subgroups. In other words, isomorphisms preserve the lattice diagram of the group. A word of caution is in order, though: the converse is not true. It is rare, but there do exist pairs of non-isomorphic groups that nonetheless have the same lattice diagram. Specifically there are two groups of order 16, one Abelian and one non-Abelian, which have the same lattice diagram. We will look at them more closely soon. The point of caution now is: y*ou can not prove that two groups are isomorphic by comparing their lattice diagrams.*

In general, this is the way you should proceed to prove two groups, G and H, are isomorphic. The first step is to make an educated guess as to what the isomorphism between the two might be. Be as specific as possible. Give a formula for the candidate isomorphism, or indicate in a function diagram how it maps elements. Be sure to define the function so that it is a bijection, and if that is not clear, prove that it is. Once you have a bijection, prove that it preserves all possible products, as in definition 30. This is not such a hard step if the function is defined by way of a formula, but if it is only defined by a diagram you must check each possible product and that can be very time-consuming. If you made a bad initial choice for the potential isomorphism, go back and make a more educated guess; see what was wrong with the first guess and correct the problem if possible. Always keep in mind that you may be trying to prove that two non-isomorphic groups are isomorphic, and how tragic is that to waste time on. There are other, more sophisticated, ways of showing that a function is an isomorphism that we will develop later.

Meanwhile, in Modern Algebra part 1 I made the claim that $\mathbf{D_3} = \mathbf{S_3}$. It should now be clear that what I meant was $\mathbf{D_3} \simeq \mathbf{S_3}$. Let's consider how to show this. First we need a bijection from $\mathbf{D_3}$ to $\mathbf{S_3}$. It is frequently best to make the simplest choice. We know the identity elements must correspond; then we will choose to map the symmetry of the triangle to the permutation of 3 letters that it naturally corresponds to once the vertices are labeled. This gives the following definition for ϕ:

$\phi(\rho_0) = (1)$ $\qquad \phi(\rho_1) = (1, 2, 3)$ $\qquad \phi(\rho_2) = (1, 3, 2)$
$\phi(\delta_1) = (2, 3)$ $\qquad \phi(\delta_2) = (1, 3)$ $\qquad \phi(\delta_3) = (1, 2)$

ϕ is certainly a bijection. The only thing that we need to do is show that it preserves products. The way we have defined ϕ requires that we check every possible product in $\mathbf{D_3}$ to verify that it maps to the corresponding product in $\mathbf{S_3}$. As one example:

$\phi(\rho_2 \cdot \delta_3) = \phi(\delta_2) = (1, 3)$ and $\phi(\rho_2) \cdot \phi(\delta_3) = (1, 2)(1, 3, 2) = (1, 3)$

where the cycles are written in the reverse order as usual. There are 25 total products to check because we need not check products with the identity element, but we do need to check products in the reverse order since the groups are non-Abelian. It is easy work but it is tedious, and you can finish checking them as an exercise. Do note that this is not the only way we could have defined ϕ, but it is the easiest way to define it. What other definition might you have tried if you weren't thinking clearly?

A simple application of the idea of isomorphisms, so simple I will not even call it a theorem, is this: *up to isomorphism, there is only one cyclic group of order n*. To see this, let G and H both be cyclic groups of order n. Then $G = < a : a^n = 1 >$ and $H = < b : b^n = 1 >$ by definition of cyclic groups. Then we define the candidate isomorphism in the most obvious possible way: $\phi(a^k) = b^k$. It is immediate to

check that this is an isomorphism; since it is defined by a formula, we need not check each possible product:
$$\phi(a^k \cdot a^m) = \phi(a^{k+m}) = b^{k+m} \quad \phi(a^k) \cdot \phi(a^m) = b^k \cdot b^m = b^{k+m}$$

When I give a list of all the groups we know, up to order 32, it should be clear that what I mean is all the groups up to isomorphism. It is of little interest to us what the names of the elements are, or what symbol we use to denote the operation. We are only interested in the algebra that we can do.

Exercise 1: Finish showing that $\mathbf{D}_3 \simeq \mathbf{S}_3$.

Exercise 2: Assume $G \simeq H$. Show that H is Abelian iff G is Abelian.

Exercise 3: Without using the result in italics above, show that $\mathbf{U}_{22} \simeq \mathbf{Z}_{10}$.

Exercise 4: Prove parts c), d), e), f) and g) of theorem 37.

Lesson 2: The Direct Product

Before we can study isomorphisms we are going to need more raw material to work with, a larger variety of groups. You should already be familiar with the Cartesian product of two sets. If S and T are any two sets, then S × T is the set of all ordered pairs (s, t) where s ϵ S and t ϵ T. There should be no cause for confusing an ordered pair with a transposition. The context should always make it perfectly clear which one is meant.

Now suppose S and T are both groups. Then it turns out there is a natural way to define a group structure on S × T. It is desirable that both S and T end up being subgroups of this new group – or rather, isomorphic to subgroups of this new group. This is all accomplished by the following. Again, for the sake of clarity, I will use additive notation for one group and multiplicative notation for the other. In the future, I won't.

Definition 31: Let $< G, +>$ and $< H, \cdot >$ be any two groups. On the *set* G × H of ordered pairs, we will define an operation * this way: $(g_1, h_1)*(g_2, h_2) = (g_1 + g_2, h_1 \cdot h_2)$. The result, $< G \times H, * >$, is called the **direct product** of G and H. G and H are called the **factors** of the direct product G × H.

I will always use the term "factor" to refer to a component of a direct product, but not all authors observe this rule. There will also be defined the term "quotient", a quite distinct idea, that I want to keep sharply separated. There are several basic facts about direct products which must be proven and which I will again collect all together into one theorem. The proofs are straight-forward and will mainly be left as exercises.

Theorem 38: Let G and H be groups and $< G \times H, *>$ their direct product. Then the following are true:
  a) $< G \times H, *>$ is a group.
  b) $G \times H \simeq H \times G$
  c) if K is another group, then $(G \times H) \times K \simeq G \times (H \times K)$
  d) $|G \times H| = |G| \cdot |H|$
  e) G × H is Abelian iff both G and H are Abelian.
  f) if **0** is the identity of G and **1** is the identity of H, then **(0, 1)** is the identity of G × H.
  g) if $(g, h) \in G \times H$ then its inverse is $(-g, h^{-1})$.
  h) if $G \simeq A \times B$, then both A and B are isomorphic to subgroups of G.
  i) if $G \simeq A \times B$ and both A and B are identified with their isomorphic subgroups, then the elements of A commute with the elements of B.
  j) the order of $(g, h) \in G \times H$ equals the least common multiple of the orders of g in G and h in H.

Proof a) take the operation defined on G as + and the operation defined on H as ·. First we must check that * is an associative operation:
  $[(g_1, h_1)*(g_2, h_2)]*(g_3, h_3) = (g_1 + g_2, h_1 \cdot h_2)*(g_3, h_3) = ((g_1 + g_2) + g_3, (h_1 \cdot h_2) \cdot h_3)$
    $= (g_1 + g_2 + g_3, h_1 \cdot h_2 \cdot h_3)$ since both + and · are associative.
  $(g_1, h_1)*[(g_2, h_2)*(g_3, h_3)] = (g_1, h_1)*(g_2 + g_3, h_2 \cdot h_3) = (g_1 + (g_2 + g_3), h_1 \cdot (h_2 \cdot h_3))$
    $= (g_1 + g_2 + g_3, h_1 \cdot h_2 \cdot h_3)$ since both + and · are associative.
Thus * is an associative operation.

Is there an identity element? A little thought shows that (0, 1) is the identity element for * as is easily checked (you will do this as an exercise for part f).

It is also easily checked that the inverse, under *, of (g, h) is the element $(-g, h^{-1})$ (and you will check it as an exercise for part g). Thus $< G \times H, * >$ is a group.

  b) We will choose a candidate isomorphism for $\phi: G \times H \to H \times G$. Making it as simple as possible, we define $\phi((g, h)) = (h, g)$. ϕ is clearly a bijection. We need only check that it preserve

products:
$\phi((g_1, h_1)(g_2, h_2)) = \phi((g_1+g_2, h_1 \cdot h_2)) = (h_1 \cdot h_2, g_1+g_2)$
$\phi((g_1, h_1)) * \phi((g_2, h_2)) = (h_1, g_1) * (h_2, g_2) = (h_1 \cdot h_2, g_1+g_2)$

Hence ϕ does preserve products and is an isomorphism.

The proofs of parts c through j are left as exercises.

A little imagination might suggest that we could take this theorem to a higher level of abstraction. What if we collect into a gigantic set all of the finite groups up to isomorphism. Is the direct product an operation on the set of all finite groups? Would it make that collection into a group itself, albeit an infinite one? Indeed, it comes close. One of the requirements for being a group is lacking to make the collection of all finite group under the operation of direct product into a group itself. Can you tell which one? What is the name for this structure?

We can nearly double the number of examples of groups we currently know by taking the ones we know in pairs and forming their direct products. We will look at a few examples of direct products in the next lesson, but in fact we have already met several direct products in part 1. Let's look at the easiest one now. We will show that $\mathbf{V} \simeq \mathbf{Z}_2 \times \mathbf{Z}_2$.

First we must name the elements of both groups. Let's call the elements of $\mathbf{V}$ {1, a, b, c} and call the elements of $\mathbf{Z}_2 \times \mathbf{Z}_2$ { (0, 0), (0, 1), (1, 0), (1, 1) }. How do we define the candidate for the isomorphism? There is no immediately obvious way to do it because all the non-identity elements are order 2; in this case the only thing that is required is to be sure to map the identity to the identity. So let's set it up as follows:

$\phi(1) = (0, 0) \qquad \phi(a) = (0, 1) \qquad \phi(b) = (1, 0) \qquad \phi(c) = (1, 1)$

We now need to check all possible products, but in this case, because the groups are Abelian we do not need to check the products when reversed; and we do not need to check products involving the identity; and, because the groups are so small, there are only three products left to check. That leaves the following:

$\phi(c) = \phi(a \cdot b) = \phi(a) \cdot \phi(b) = (0, 1) * (1, 0) = (0+1, 1+0) = (1, 1)$ which checks.
$\phi(b) = \phi(a \cdot c) = \phi(a) \cdot \phi(c) = (0, 1) * (1, 1) = (0+1, 1+1) = (1, 0)$ which checks.
$\phi(a) = \phi(c \cdot b) = \phi(c) \cdot \phi(b) = (1, 1) * (1, 0) = (1+1, 1+0) = (0, 1)$ which checks.

Since we are beginning with the Cartesian product of sets, we know that two ordered pairs are equal iff their components are equal: (a, b) = (c, d) iff a = c and b = d. To take advantage of this we must remember that all the algebra in a direct product is done *component-wise*. We do the algebra in G and in H separately and then put them together in their components.

Exercise 5: Prove theorem 38 parts b), c), d), e), f), g), h), i) and j).
Exercise 6: Prove that both G and H are isomorphic to subgroups of G × H.
Exercise 7: Let $G \simeq A \times B$ and $H \simeq B \times C$ be groups. Show that $G \times C \simeq A \times H$.

Lesson 3: Direct Products of Cyclic Groups

The purpose of this lesson is to examine some of the simplest direct products in some detail. To be manageable we must focus on products of relatively small groups. First though, we will prove a theorem that will make some direct products very easy to compute. We will now dispense with the practice of symbolizing the different operations in the groups by different symbols. I will use multiplicative notation for the cyclic groups in the next theorem, and then switch to additive notation in the examples.

Theorem 39: Let m and n be relatively prime integers. Then $Z_m \times Z_n \simeq Z_{mn}$.

Proof: Let $Z_m = <a>$ and let $Z_n = <b>$. In $Z_m \times Z_n$ consider the element (a, b). Since m and n are relatively prime, k = mn is the least common multiple and so (a, b) has order km by theorem 38j. Since the order of $Z_m \times Z_n$ equals k, (a, b) generates all of $Z_m \times Z_n$. Therefore $Z_m \times Z_n \simeq Z_{mn}$.

This tells us a lot of trivial direct products:
$$Z_2 \times Z_3 \simeq Z_6$$
$$Z_8 \times Z_7 \simeq Z_{56}$$
$$Z_{101} \times Z_{40} \simeq Z_{4040}$$

and so on. To get more complicated direct products we must choose two cyclic groups whose orders are not relatively prime. We then examine the elements individually to determine what subgroups they generate, and follow the same procedure we used in part 1.

This theorem also gives us a preferred way of writing the direct products of cyclic groups that is somewhat standardized. It is always helpful to know the element that has the largest possible order in the group, then the next largest order not already included, and so on. If the group is not already written in this form, first split each given factor into a product of as many relatively prime cyclic factors as possible. Then choose the largest factors involving each available prime and put them together as the largest possible factor. Repeat the process with the remaining available factors until all are accounted for. The advantage of this factorization is that it reveals the existence of the elements with the largest orders, making the overall structure of the group as visible as possible.

As an example of this preferred way of writing products, consider the following:
$$Z_{100} \times Z_{40} \times Z_{70} \simeq (Z_4 \times Z_{25}) \times (Z_8 \times Z_5) \times (Z_2 \times Z_5 \times Z_7)$$
$$\simeq (Z_8 \times Z_{25} \times Z_7) \times (Z_4 \times Z_5) \times (Z_2 \times Z_5)$$
$$\simeq Z_{1400} \times Z_{20} \times Z_{10}$$

Without rewriting the product in this more convenient form, we would not have noticed so quickly that there is an element of order 1400. Knowing this fact makes it much easier to understand what is going on inside the product. It is usually best to rewrite a direct product of cyclic groups in this form as a first step in working with them.

As our first example, let's consider $Z_2 \times Z_4$. Note them I am not using the standard form for this product. It is small enough that I have not been careful about the order of the factors. The elements are:

(0, 0)   (0, 1)   (0, 2)   (0, 3)   (1, 0)   (1, 1)   (1, 2)   (1, 3)

We should examine the elements of the highest order first to get the most accomplished right off the bat. The order of an element is the least common multiple of the orders of the components, so it is easy to see that four of these elements are order 4 and that is the maximal order; with two elements of order 4 for each copy of Z_4 we expect there to be two copies of Z_4. It is easy to calculate, first the factor we begin with and then the new copy:

(0, 1) + (0, 1) = (0, 2)   (0, 2) + (0, 1) = (0, 3)   (0, 3) + (0, 1) = (0, 0)   < (0, 1) >
(1, 1) + (1, 1) = (0, 2)   (0, 2) + (1, 1) = (1, 3)   (1, 3) + (1, 1) = (0, 0)   < (1, 1) >

These copies of Z_4 share a common Z_2 and account for six of the elements. The other two elements are both order 2, making three elements of order 2: (0, 2), (1, 2) and (1, 0). It is easy to check that these three form a copy of the Klein 4-group **V**. These are all the subgroups. In the past we have been casual about asserting that we have found all the subgroups; we will shortly prove the theorem that assures us our casual assumption has been valid.

Note that there is a kind of distributive rule for direct products. Each subgroup of the first factor forms a direct product with each subgroup of the second factor. We will see this more explicitly in our next example, $Z_2 \times Z_8$. Z_8 contains a single copy of Z_4 and a single copy of Z_2 as subgroups, so we should expect to find copies of $Z_2 \times Z_4$ and $Z_2 \times Z_2$ in the product $Z_2 \times Z_8$.

The elements of $Z_2 \times Z_8$ are the following:

(0, 0)　(0, 1)　(0, 2)　(0, 3)　(0, 4)　(0, 5)　(0, 6)　(0, 7)
(1, 0)　(1, 1)　(1, 2)　(1, 3)　(1, 4)　(1, 5)　(1, 6)　(1, 7)

We see that there are eight elements that have order 8, and since four are required for each copy of Z_8 there are two such copies as listed below with the factor group listed first:

(0, 1) → (0, 2) → (0, 3) → (0, 4) → (0, 5) → (0, 6) → (0, 7) → (0, 0)
(1, 1) → (0, 2) → (1, 3) → (0, 4) → (1, 5) → (0, 6) → (1, 7) → (0, 0)

They share their copy of Z_4 so this accounts for twelve out of the sixteen elements. Of the remaining four, two are order 4 – namely (1, 2) and (1, 6) – so these determine a single copy of Z_4 which shares the element of order 2 with the two copies of Z_8. The remaining elements are (1, 4) and (1, 0) neither of which is in another cyclic subgroup. Together the three elements of order 2, (0, 4), (1, 4), and (1, 0), form a copy of **V**. The lattice diagrams for these two direct products are given at the end of the lesson. This is the Abelian group whose lattice diagram is identical to the lattice diagram of a non-Abelian group called the **Modular group**, **M**. In the exercises you will be asked to check this fact.

We now turn to the presentation of direct products. As you might expect the set of generators of a direct product is the union of the sets of generators of the factors. The relations must include all the relations of the presentations of the factors, but in addition we must include relations for how the generators of the different factors are related. It should be clear that any generator in one factor must commute with any generator in another factor, and that is what is necessary to state in the presentation.

If we are forming the direct product of cyclic groups then there is only one generator for each factor and the task is easy. The presentation of $Z_4 \times Z_4$ is just $<a, b : a^4 = b^4 = 1, ab = ba>$. If there are more that two factors, we must be careful to include relations between all pairs of generators, and of course the presentation we give will depend on the factors that are explicitly involved in the way the product is written. Thus

$<a, b, c : a^{600} = b^{20} = c^{20} = 1, ab = ba, ac = ca, bc = cb> \simeq Z_{600} \times Z_{20} \times Z_{20}$
$\simeq Z_{100} \times Z_{40} \times Z_{60} \simeq <a, b, c : a^{100} = b^{40} = c^{60} = 1, ab = ba, ac = ca, bc = cb>$

The Cayley digraph is constructed as usual.

Exercise 8: Give the presentation and the Cayley digraph for $Z_2 \times Z_4$ and $Z_2 \times Z_8$. The modular group has the presentation $< g, h : g^2 = h^8 = 1, hg = gh^5 >$. Show that the modular group **M** has the same lattice diagram as $Z_2 \times Z_8$.

Exercise 9: Prove $Z_2 \times Z_4 \simeq U_{15}$. Give the isomorphism explicitly. This result will be used in an important exercise in the next lesson.

Exercise 10: Construct the complete lattice diagram and the Cayley digraph for $Z_4 \times Z_4$. Give a presentation for it. (This result will be used in the next set of exercises.)

Exercise 11: Prove that $Z_2 \times Z_6 \simeq U_{21}$. Give the isomorphism explicitly.

Exercise 12: Prove that $U_{52} \simeq U_{70}$.

Exercise 13: Express the following direct products in the standard form:
$Z_{250} \times Z_{75} \times Z_{96}$ $Z_{48} \times Z_{48} \times Z_{240}$ $Z_{90} \times Z_{81} \times Z_{250}$ $Z_{60} \times Z_{65} \times Z_{110} \times Z_{72}$

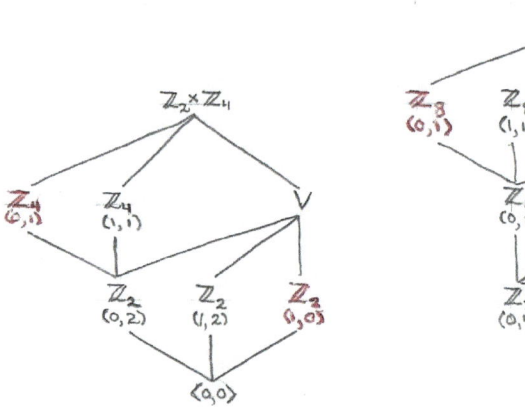
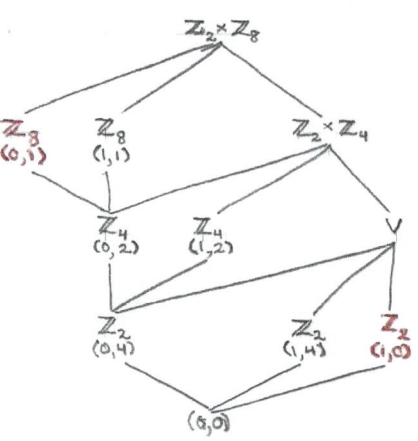

Lesson 4: More Complicated Direct Products

As two final examples, let's first consider $\mathbf{Z}_2 \times \mathbf{Z}_2 \times \mathbf{Z}_6$. Note that I have again not bothered to write this product in its standard form because nothing is gained in doing it. Its elements are triples, which I now list:

(0, 0, 0)	(0, 0, 1)	(0, 0, 2)	(0, 0, 3)	(0, 0, 4)	(0, 0, 5)
(0, 1, 0)	(0, 1, 1)	(0, 1, 2)	(0, 1, 3)	(0, 1, 4)	(0, 1, 5)
(1, 0, 0)	(1, 0, 1)	(1, 0, 2)	(1, 0, 3)	(1, 0, 4)	(1, 0, 5)
(1, 1, 0)	(1, 1, 1)	(1, 1, 2)	(1, 1, 3)	(1, 1, 4)	(1, 1, 5)

The three factors of the product account for eight out of the twenty-four elements. It is most efficient to find other elements of order 6 first and determine the subgroups they generate. There are twelve elements of order 6 which will determine six new copies of $\mathbf{Z}_6$.

(0, 1, 1) + (0, 1, 1) = (0, 0, 2)	(1, 0, 1) + (1, 0, 1) = (0, 0, 2)	(1, 1, 1) + (1, 1, 1) = (0, 0, 2)
(0, 0, 2) + (0, 1, 1) = (0, 1, 3)	(0, 0, 2) + (1, 0, 1) = (1, 0, 3)	(0, 0, 2) + (1, 1, 1) = (1, 1, 3)
(0, 1, 3) + (0, 1, 1) = (0, 0, 4)	(1, 0, 3) + (1, 0, 1) = (0, 0, 4)	(1, 1, 3) + (1, 1, 1) = (0, 0, 4)
(0, 0, 4) + (0, 1, 1) = (0, 1, 5)	(0, 0, 4) + (1, 0, 1) = (1, 0, 5)	(0, 0, 4) + (1, 1, 1) = (1, 1, 5)
(0, 1, 5) + (0, 1, 1) = (0, 0, 0)	(1, 0, 5) + (1, 0, 1) = (0, 0, 0)	(1, 1, 5) + (1, 1, 1) = (0, 0, 0)
(0, 1, 2) + (0, 1, 2) = (0, 0, 4)	(1, 0, 2) + (1, 0, 2) = (0, 0, 4)	(1, 1, 2) + (1, 1, 2) = (0, 0, 4)
(0, 0, 4) + (0, 1, 2) = (0, 1, 0)	(0, 0, 4) + (1, 0, 2) = (1, 0, 0)	(0, 0, 4) + (1, 1, 2) = (1, 1, 0)
(0, 1, 0) + (0, 1, 2) = (0, 0, 2)	(1, 0, 0) + (1, 0, 2) = (0, 0, 2)	(1, 1, 0) + (1, 1, 2) = (0, 0, 2)
(0, 0, 2) + (0, 1, 2) = (0, 1, 4)	(0, 0, 2) + (1, 0, 2) = (1, 0, 4)	(0, 0, 2) + (1, 1, 2) = (1, 1, 4)
(0, 1, 4) + (0, 1, 2) = (0, 0, 0)	(1, 0, 4) + (1, 0, 2) = (0, 0, 0)	(1, 1, 4) + (1, 1, 2) = (0, 0, 0)

All of these copies of $\mathbf{Z}_6$ share the single copy of $\mathbf{Z}_3$ from the third factor of the product, but each with its own copy of $\mathbf{Z}_2$. The third and fourth copies of $\mathbf{Z}_6$ listed above incorporate the two $\mathbf{Z}_2$'s from the first two factors of the product. We have now accounted for all the elements.

So are there any copies of $\mathbf{V}$? At this point it is critical to be systematic. It takes three elements of order 2 to produce a copy of $\mathbf{V}$. No two copies of $\mathbf{V}$ can share more than one element since two elements determine the entire subgroup. If the product of a pair of elements equals another element of order 2, then those three elements determine a copy of $\mathbf{V}$. Let's begin enumerating them by focusing on (1, 0, 0):

(1, 0, 0) + (0, 1, 0) = (1, 1, 0)

and so these three elements form a copy of $\mathbf{V}$. We next try putting (1, 0, 0) with one of the four remaining order 2 elements:

(1, 0, 0) + (1, 1, 3) = (0, 1, 3)
(1, 0, 0) + (1, 0, 3) = (0, 0, 3)

Continuing in this way, trying different pairs of order 2 elements, and discarding duplicates, we can find four more copies of $\mathbf{V}$:

(0, 1, 0) + (0, 1, 3) = (0, 0, 3)
(0, 1, 0) + (1, 1, 3) = (1, 0, 3)
(1, 1, 0) + (0, 1, 3) = (1, 0, 3)
(1, 1, 0) + (1, 1, 3) = (0, 0, 3)

This is quite the most complicated group structure we have tried to diagram in detail and the lattice diagram is difficult to get right. I have given my best stab at it. Again, the subgroups corresponding to the original factors are shown in red.

However, the lattice diagram I am giving here leaves out several subgroups. Recall that there is a distributive pattern in the product. Each subgroup of each factor produces at least one product with each subgroup of the other factors. Thus we expect, in addition to what is shown here, some subgroups that are isomorphic to $Z_2 \times Z_6$. In exercise 11 of the previous lesson, you showed that the product $Z_2 \times Z_6$ is isomorphic to U_{21}, whose lattice diagram was determine in lesson 12 of part 1. It is an exercise for you to find all the copies of U_{21} that

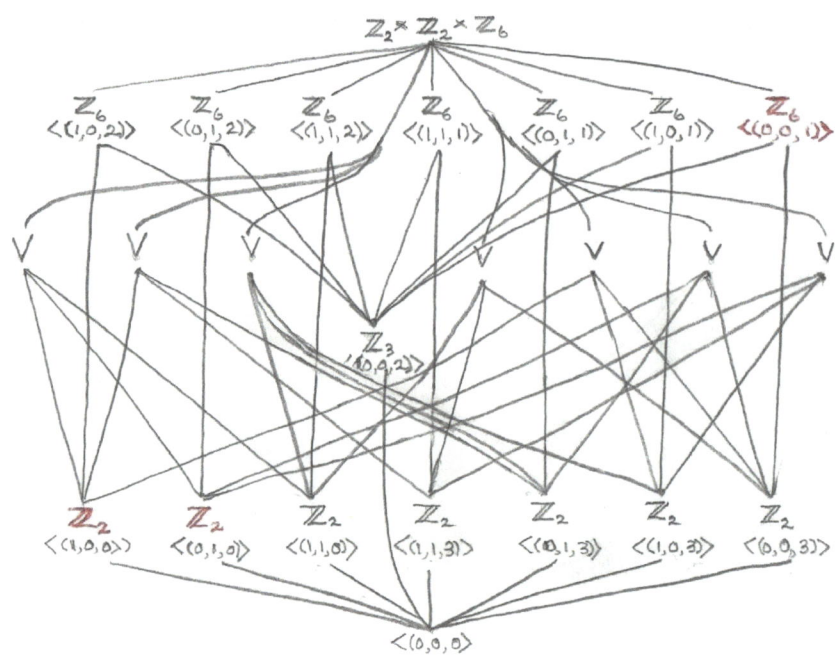

are present and how they are connected to the copies of Z_6. Do not make the rash assumption that there are only two such copies.

The Cayley digraph of $Z_2 \times Z_2 \times Z_6$ is also shown. Since there are three generators for this group, I have used three colors in the diagram. The Cayley digraphs of direct products of cyclic groups have a signature pattern, if you arrange them right, that we should have expected with some fore thought.

The next product we will consider in some detail is a non-Abelian one. If either factor of the product is non-Abelian, then the product is as well. Non-Abelian products are more complicated in structure, as you should expect. The smallest direct product in which both factors are non-Abelian is $D_3 \times D_3$ which has order 36. It takes a lot of work and is quite a tangle of lines when you are done with it. It is not an exercise. You could consider it a challenge to your skill.

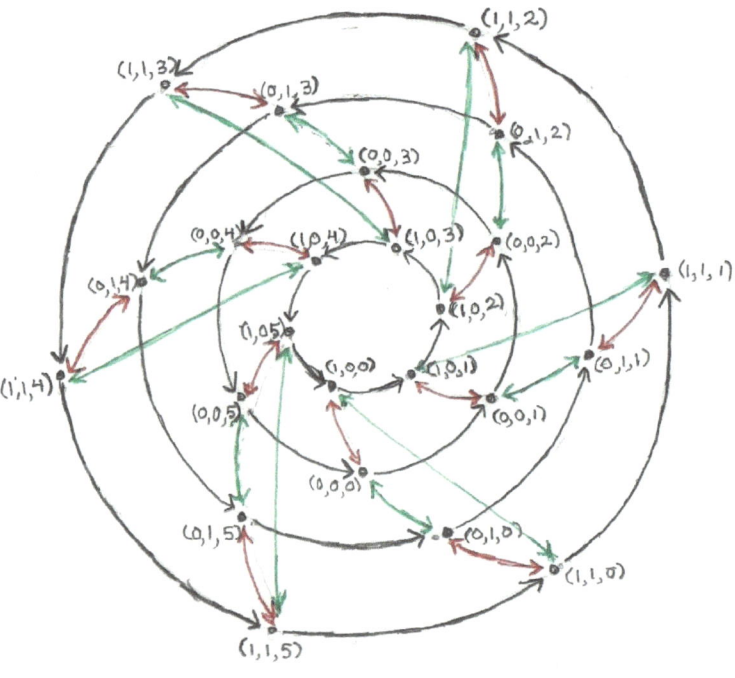

Instead we will consider $Z_3 \times D_4$. The subgroups of D_4 are Z_4, Z_2 and V. Hence we expect the product to contain subgroups isomorphic to $Z_3 \times Z_4$ (which is isomorphic to Z_{12} by theorem 39), $Z_3 \times Z_2$ (isomorphic to Z_6), and $Z_3 \times V$ (which is isomorphic to $Z_6 \times Z_2$). Further, since there will be a product of $\mathbf{Z}_3$ with each of the five copies of $\mathbf{Z}_2$ contained in $\mathbf{D}_4$, we expect at least five new copies of $\mathbf{Z}_6$ to be produced. The verification of my diagram I will include as part of the exercise, but I have indicated the generators of the cyclic subgroups to give you a start. I again leave out the subgroups isomorphic to $\mathbf{U}_{21}$ for you to discover for yourself. This lattice diagram is simpler than you might have expected. Since the first factor has no element of order 2 we do not get a proliferation of elements of order 2 in the product. In fact, since the orders of the elements in the first factor are relatively prime to the orders of all the elements in the second factor, most of the "cross" products are simple cyclic subgroups.

There is no standard order for writing the factors of direct products involving non-Abelian groups. I would prefer to put cyclic factors first. One exercise that is important as a counter-example in a future lesson is exercise 15. It is quite an involved lattice diagram but worth spending the time on.

Exercise 14: Find the missing subgroups in the lattice diagram for $\mathbf{Z}_2 \times \mathbf{Z}_2 \times \mathbf{Z}_6$ explicitly by giving the elements and redrawing the lattice diagram with every subgroup included.

Exercise 15: Compute the structure of $\mathbf{Z}_3 \times \mathbf{D}_4$ in detail using the lattice diagram given below as a guide. Be sure to include all the subgroups. Then draw a Cayley digraph for this group.

Exercise 16: Completely determine the structure of $\mathbf{Z}_4 \times \mathbf{Q}_8$. We have already determined the structures of all the subgroups that will be found in this product (if you did exercise 11) so the only issues will be how many of each there are, and how they are inter-connected. This will provide us with a counter-example in a future lesson.

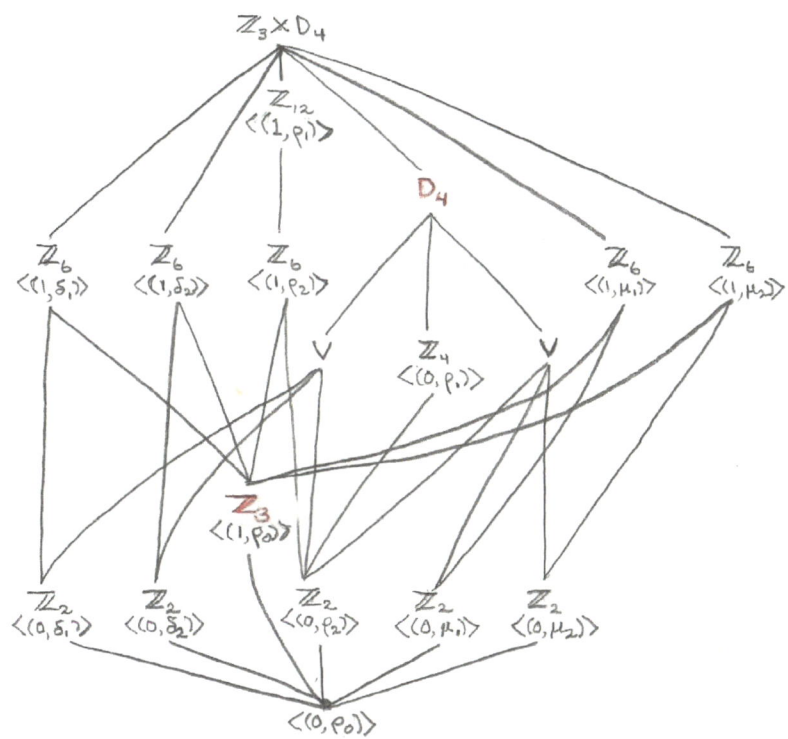

Lesson 5: Three Theorems about Direct Products

Before we go on to other material, there are several loose ends that should be tied up, and two important results we will frequently need to use. The idea of the direct product enables us to condense our knowledge to this point into fewer and tighter packages. The next theorem ties up a loose end and is surprisingly easy to prove:

Theorem 40: Let m be an odd positive integer and n = 2m. Then $\mathbf{D_n} \simeq \mathbf{Z_2} \times \mathbf{D_m}$

Proof: In order to show that $\mathbf{D_n} \simeq \mathbf{Z_2} \times \mathbf{D_m}$ we must find two order 2 elements of $\mathbf{Z_2} \times \mathbf{D_m}$ whose product is an element of order n. Then we would know that those two elements together generate $\mathbf{D_n}$. Since $|\mathbf{D_n}| = |\mathbf{Z_2} \times \mathbf{D_m}|$ we will know the two groups must be the same. So, let $\mathbf{Z_2} = <a>$ and let $\mathbf{Z_2} \times \mathbf{D_m} = <\delta_1, \rho>$. We use δ's for the reflections since m is odd and there won't be any reflections of the μ type. As to rotations, we only require that ρ be of order m. There are no rotations of order 2 in $\mathbf{D_m}$ because m is odd.

We know that in $\mathbf{D_m}$ there are two elements of order 2, say δ_1 and δ_2, such that $\delta_1 \cdot \delta_2 = \rho$, the generator of order m (see part 1, lesson 28). Now consider the two ordered pairs (a, δ_1) and (1, δ_2). They are both order 2 in $\mathbf{Z_2} \times \mathbf{D_m}$, and their product is (a, ρ) which has order n, the least common multiple of 2 and m. Hence < (a, δ_1), (1, δ_2) > = $\mathbf{D_n}$, which must be the entirety of $\mathbf{Z_2} \times \mathbf{D_m}$.

So far we have considered direct products from the viewpoint of building the product out of two groups we have previously chosen. It is much harder to reverse the process. What if we take an arbitrary group and ask whether that group is, in fact, isomorphic to the direct product of two of its subgroups. We think of this process, of going down from the whole group to a direct product of two or more of its subgroups, as an **internal direct product**. We wish to have a test that can determine whether a particular group is an internal direct product; or, in other words, whether a given group be factored. It can be a difficult question to answer, but the next theorem gives considerable help, though only for finite Abelian groups. The factoring of non-Abelian groups requires a few more tools and we will return to that problem later.

Theorem 41: Let G be a finite Abelian group; let H and K be two subgroups of G. Then $G \simeq H \times K$ if and only if $H \cap K = 1$ and $H \vee K = G$.

Proof: ⇒) First suppose G is Abelian and $G \simeq H \times K$. Then $H \simeq \{(h, 1)\}$ and $K \simeq \{(1, k)\}$. Hence we see $H \cap K = (1, 1)$ and it is also clear that $H \vee K = G$.

⇐) Now suppose $H \cap K = 1$ and $H \vee K = G$. We will define a candidate isomorphism this way: φ: H × K → G by φ((h, k)) = hk. It is easy to see that φ is a well-defined function; no ordered pair is mapped to two distinct products in G. Furthermore, since these are finite groups, we know that φ is a bijection if it is either an injection or a surjection. Suppose φ is not an injection. Then assume φ((h_1, k_1)) = φ((h_2, k_2)). If this is true, then $h_1 k_1 = h_2 k_2$ by definition of φ. Hence $h_2^{-1} h_1 = k_2 k_1^{-1}$. But since $H \cap K = 1$ we see that $h_2^{-1} h_1 = k_2 k_1^{-1} = 1$, and therefore $h_2 = h_1$ and $k_2 = k_1$. Hence φ is an injection and therefore a bijection. We need only show that it preserves products:

φ((h_1, k_1)·(h_2, k_2)) = φ(($h_1 \cdot h_2$, $k_1 \cdot k_2$)) = $h_1 \cdot h_2 \cdot k_1 \cdot k_2$ = $h_1 k_1 \cdot h_2 k_2$ (since G is Abelian)
= φ((h_1, k_1))·φ((h_2, k_2))

Therefore $G \simeq H \times K$.

Corollary 41a: Let G be a finite Abelian group and suppose $\mathbf{Z_n} \leq G$, $\mathbf{Z_m} \leq G$, with n and m being relatively prime. Then $\mathbf{Z_n} \times \mathbf{Z_m} \leq G$.

Proof: is left as an exercise.

In combination with theorem 41, the next theorem is a very powerful one. Together they tells us the structure of every finite Abelian group. It is a theoretical accomplishment more than a practical one as finiteness always becomes too large eventually. There is no such sweeping theorem for finite non-Abelian groups, which are just intrinsically more complicated and varied in structure. We will be able to extend theorem 41 to non-Abelian groups however, in lesson 13, and that will be quite useful then.

Theorem 42: Let G be a finite Abelian group. Then G is isomorphic to a direct product of cyclic groups.

Proof: First note that since the group operation is commutative, the distributive law of exponents over multiplication holds in both directions; specifically $(a \cdot b)^n = a^n \cdot b^n$.

We next prove a fact we will need in passing. This sort of fact is usually called a *lemma*. This one anticipates a theorem that we will prove in a more general setting in part 3.

Lemma 42a: Let $|G| = n = p_1^{m_1} p_2^{m_2} p_3^{m_3} \ldots p_k^{m_k}$ where the p_i are all the distinct prime divisors of n. Let $P_i = \{ g \in G \mid$ the order of g is a power of $p_i \}$. Clearly $1 \in P_i$. We know the order of the inverse of an element equals the order of the element, so P_i includes the inverses of all its elements. Now let $g, h \in P_i$. If g and h are both powers of the same element of P_i then clearly $gh \in P_i$. So we may assume g and h cannot be expressed as powers of a single element. The law of exponents tells us that $(gh)^m = g^m h^m = \mathbf{1}$ iff either $g^m = h^{-m}$ (contradicting our assumption), or else $g^m = h^m = \mathbf{1}$. Thus m must be a power of p_i and $gh \in P_i$. Hence $P_i \leq G$.

We now have a collection of subgroups of G, namely $P_1, P_2, \ldots, P_k$. It is clear that whenever $i \neq j$ then $P_i \cap P_j = 1$; otherwise there would be elements whose orders equalled powers of two distinct primes. Now consider $P_2 \vee P_3 \vee \ldots \vee P_k = C_1$. It is clear that $P_1 \cap C_1 = \mathbf{1}$. Further, it is easy to see that $P_1 \vee C_1 = G$. Therefore, by theorem 41, we conclude $G \simeq P_1 \times C_1$.

Now $|C_1| = p_2^{m_2} p_3^{m_3} \ldots p_k^{m_k}$ and we will repeat the process from the previous paragraph. We have $P_2 \leq C_1$, and we will form $C_2 = P_3 \vee \ldots \vee P_k$ and then conclude that $C_1 \simeq P_2 \times C_2$. There is nothing to stop us from continuing this process, which must terminate after a finite number of steps, with $G \simeq P_1 \times P_2 \times P_3 \ldots \times P_k$.

However the P_i are not necessarily cyclic. We must breakdown our factorization of G into smaller factors and we will use a similar method to do it. Let $q \in P_i$ be an element of maximum order in P_i. If $<q> = P_i$ we are done, so let's assume the order of q is p_i^t, and that $t < m_i$. Then $\mathbf{Z_q} \leq P_i$. Now let r be an element of maximal order in P_i that is not in $<q>$, say of order p_i^s. Then we have $<r> \cap <q> = 1$. $<r> \vee <q>$ may not equal P_i but it is certainly a subgroup of P_i and an Abelian group in its own right. Therefore, by corollary 41a,
$$<r> \vee <q> \simeq <r> \times <q> \leq P_i$$
Both factors are cyclic. If it is still the case that $<r> \times <q>$ is not isomorphic to P_i, choose another element of maximal order that is in P_i but not in $<r> \times <q>$ and use the same technique to get a larger direct product. Since P_i is a finite group, the process must end with an expression for P_i as a direct product of cyclic groups. Now insert these factorizations for each of the P_i into the product $G \simeq P_1 \times P_2 \times P_3 \ldots \times P_k$ and we get the factorization of G into a direct product of cyclic groups.

Hence, we effectively know all the finite Abelian groups. The infinite Abelian groups are quite a different matter, of course, and we will discuss them in part 5 of this series.

Exercise 17: List all of the Abelian groups up to order 32.

Exercise 18: Prove corollary 41a.

Exercise 19: Can theorem 40 be generalized in this way: let p be a prime, let m be a positive integer not divisible by p, and let n = pm. Is it always true that $\mathbf{D_n} \simeq \mathbf{Z_p} \times \mathbf{D_m}$? To see if this conjecture is plausible, explicitly check whether $\mathbf{D_{12}} \simeq \mathbf{Z_3} \times \mathbf{D_4}$ by construction.

Lesson 6: U_n and the Chinese Remainder Theorem

Now we know that all finite Abelian groups are either cyclic or direct products of cyclic groups. Hence we also know that U_n are finite Abelian groups, and so they can all be expressed as direct products of cyclic groups. We know that if n equals 2, 4, p^k, or $2p^k$ (where p is an odd prime) then n has a primitive root; that is, U_n has a single generator and so is cyclic and isomorphic to $Z_{\varphi(n)}$. What we don't know is this: what product of cyclic groups is U_n isomorphic to when n does not have a primitive root? The principle theorem that answers the question is the next one.

Theorem 43: Let n = mk where m and k are relatively prime. Then $U_n \simeq U_m \times U_k$.

Proof: To prove this theorem we will construct a specific isomorphism that yields the required relationship. Choose $x \in U_n$ and define $\phi: U_n \to U_m \times U_k$ by $\phi(x) = (x \pmod{m}, x \pmod{k})$.

ϕ is clearly a well-defined function; each choice for x leads to a unique ordered pair. It is also easy to see that ϕ is an injection:

$\phi(x) = \phi(y)$ iff $(x \pmod{m}, x \pmod{k}) = (y \pmod{m}, y \pmod{k})$
                iff $x \pmod{m} = y \pmod{m}$ and $x \pmod{k} = y \pmod{k}$
                iff $x = y + i \cdot m$ and $x = y + j \cdot k$ for some positive integers i and j
                iff $y + i \cdot m = y + j \cdot k$
                iff $i \cdot m = j \cdot k$ for some positive integers i and j
                iff $i = i'k$ and $j = j'm$ since k and m are relatively prime
                iff $i'km = j'mk$
                iff $i' = j'$.

Hence ϕ is injective and since these are finite groups, ϕ is a bijection.

It only remains to show that ϕ preserves products:

$\phi(xy) = ((xy \pmod{m}, xy \pmod{k}))$
        $= ((x \pmod{m}, x \pmod{k})) \cdot ((y \pmod{m}, y \pmod{k}))$
        $= \phi(x) \cdot \phi(y)$

Therefore ϕ is an isomorphism.

Thus, to find U_n as a direct product of cyclic groups, we first find the prime factorization of n: say $n = p_1^{a_1} p_2^{a_2} \ldots p_k^{a_k}$. Then

$$U_n \simeq U_{p_1^{a_1}} \times U_{p_2^{a_2}} \times \ldots \times U_{p_k^{a_k}} \simeq Z_{\varphi(p_1^{a_1})} \times Z_{\varphi(p_2^{a_2})} \times \ldots \times Z_{\varphi(p_k^{a_k})}$$

Let's do a few examples; they are all easy enough.

$U_{21} \simeq U_7 \times U_3 \simeq Z_{\varphi(7)} \times Z_{\varphi(3)} \simeq U_6 \times U_2$ confirming what we already knew.

$U_{891} \simeq U_{81} \times U_{11} \simeq Z_{\varphi(3^4)} \times Z_{\varphi(11)} \simeq Z_{2 \cdot 3^3} \times Z_{10} \simeq Z_{54} \times Z_{10} \simeq Z_{270} \times Z_2$

where the last factorization is in standard form.

There is only one catch. We know that if k > 2 then 2^k does not have a primitive root and U_{2^k} is not cyclic. Still U_{2^k} is itself a product of cyclic groups though theorem 43 gives us no help in finding it. If 2 is one of the primes in the factorization of n and has an exponent greater than 2, we do not yet know how to express U_n as a direct product. We must prove a separate theorem for this case.

Theorem 44: $U_{2^k} \simeq Z_2 \times Z_{2^{(k-2)}}$ for k > 2.

Proof: To prove this, we will show that when k > 3, $5 \in U_{2^k}$ has order 2^{k-2}. To get this we will first use the binomial theorem:

$5^{2^{(k-2)}} = (1 + 2^2)^{2^{(k-2)}} = 1 + {}_{2^{(k-2)}}C_1(2^2) + {}_{2^{(k-2)}}C_2 \cdot (2^2)^2 + \ldots + {}_{2^{(k-2)}}C_{2^{(k-2)}} \cdot (2^2)^{2^{(k-2)}}$
           $= 1 + 2^{k-2} \cdot 2^2 + [2^{k-2} \cdot (2^{k-2} - 1)/2](2^2)^2 + \ldots + 2^{2^{(k-1)}}$
           $= 1 + 2^k + 2^k \cdot 2 \cdot (2^{k-2} - 1) + \ldots + 2^k \cdot 2^{2^{(k-1)} - k}$ ($2^{k-1} - k > 0$ if k > 3)
           $= 1 + 2^k \cdot [2 \cdot (2^{k-2} - 1) + \ldots + 2^{2^{(k-1)} - k}] \equiv 1 \pmod{2^k}$

This shows that the order of 5 is at most 2^{k-2}. To see that the order of 5 is not smaller than 2^{k-2} we will see if it is perhaps order 2^{k-3}:

$$5^{2^{\wedge}(k-3)} = (1 + 2^2)^{2^{\wedge}(k-3)} = 1 + {}_{2^{\wedge}(k-3)}C_1(2^2) + {}_{2^{\wedge}(k-3)}C_2 \cdot (2^2)^2 + \ldots + {}_{2^{\wedge}(k-3)}C_{2^{\wedge}(k-3)} \cdot (2^2)^{2^{\wedge}(k-3)}$$
$$= 1 + 2^{k-1} + \ldots \not\equiv 1 \pmod{2^k}$$

Since we have never proven than 2^k for $k > 3$ has no primitive root, the proof is still incomplete. Perhaps 5 is merely a generator of the unique subgroup of order 2^{k-1} in the cyclic group of order 2^k. In the exercises you will be asked to find three distinct elements of $U_{2^{\wedge}k}$ which have order 2. This will show that $U_{2^{\wedge}k}$ is not cyclic because a cyclic group has only one subgroup of each order. This will complete the proof that

$$U_{2^{\wedge}k} \simeq <5> \times Z_2 \simeq Z_{2^{\wedge}(k-2)} \times Z_2.$$

The last theorem of this lesson is one of the oldest theorems in mathematics, called the Chinese Remainder Theorem. The proof will be left as an exercise in the use of these isomorphism we have just constructed.

Theorem 45: (The Chinese Remainder Theorem) Let $n_1, n_2, \ldots n_k$ be positive integers that are pairwise relatively prime (i. e. any two of them are relatively prime to each other), and let $a_1, a_2, \ldots a_k$ be any integers. Then there is a simultaneous solution to the system of congruences

$x \equiv a_1 \pmod{n_1}$
$x \equiv a_2 \pmod{n_2}$
$\vdots$
$x \equiv a_k \pmod{n_k}$

which is unique mod $n_1 n_2 n_3 \cdots n_k$.

Proof: is an exercise.

The introduction of the direct product has now enabled us to simplify the organization of our knowledge of finite groups. We know that all finite Abelian groups are either cyclic or direct products of cyclic groups. We know that all the groups previously labeled U_n that seemed to have no pattern to their character actually do have a systematic pattern as direct products of cyclic groups. We know that some dihedral groups are the same as direct products of cyclic groups with smaller dihedral groups. There is a great deal more to be said about the direct products of groups, but to say it we need a more elaborate conceptual framework. We next return to the study of isomorphisms and specialize them in a surprising way.

Exercise 20: Express U_{48}, U_{648}, U_{3072}, and U_{19712} as direct products of cyclic groups:.
Exercise 21: Prove the Chinese Remainder Theorem.
Exercise 22: Solve the system of congruences $x \equiv 3 \pmod 8$, $x \equiv 2 \pmod{25}$, $x \equiv 1 \pmod{81}$.

Lesson 7: Automorphisms

The best way to make progress in our study of groups is to simplify a concept that already may appear to be as simple as possible. The isomorphism has been quite useful for determining when two seemingly distinct groups are actually identical in structure. We will now simplify this nearly to the point of absurdity. We already know that a group G is identical in structure to itself. Of what use can it be to study the isomorphism from G onto itself? Yet that is exactly the right thing to do.

Definition 32: Let $\phi: G \to G$ be an isomorphism. In this case we will call ϕ an **automorphism**.

Now, an automorphism is a bijection of a set onto itself by definition. Therefore by definition an automorphism is a permutation of the elements of G. But an automorphism is not just any permutation of the elements of a group; it is a special permutation that also perfectly preserves the algebra of G. It has all the properties of an isomorphism. It must map the identity element of G to itself. It must map any element of G to another element of the same order. It must map the inverse of an element to the inverse of the image of that element. If $H \leq G$, then $\phi(H) \leq G$, and further H must be isomorphic to $\phi(H)$.

In fact there are so many restrictions on what an automorphism is allowed to do that we may initially wonder if there are any non-trivial automorphisms at all – for of course the identity map is clearly an automorphism. To investigate this, we naturally look at the simplest of groups, say Z_3. You can convince yourself easily that the even more simple group Z_2 has no non-trivial automorphisms. Is there a non-trivial automorphism of Z_3? Before we try to find it, it is worth recalling that we need only consider how the automorphism maps the generators of the group in question because the automorphism must preserve all the algebra created by the generators. Since a cyclic group has only one generator, these will be the easiest groups to consider.

The elements of Z_3 in additive notation are 0, 1, and 2. We already know that any automorphism must map 0 to 0. We take 1 as the generator of Z_3 and we seem to have two choices of where to map 1. If 1 is mapped to 1, then we are forced to map 2 to 2 and we end up with the identity map. Can 1 be mapped to 2?

$$\text{If } \phi(1) = 2, \text{ then } \phi(2) = \phi(1 + 1) = \phi(1) + \phi(1) = 2 + 2 = 1$$

Thus we have an automorphism that fixes 0 and switches 1 and 2. Note that under composition ϕ^2 is just the identity map. Since ϕ is a permutation of Z_3 we can write ϕ in cycle notation this way: $\phi = (1, 2)$. There are no other possibilities for the images of 1 and so get only these two automorphisms of Z_3.

Let's up the game a bit by finding all the automorphisms of Z_4. Again using additive notation, the elements of Z_4 are 0, 1, 2, and 3. Again 0 must be mapped to 0. We will take 1 as the generator and immediately notice that 1 can be mapped to 3 but 1 cannot be mapped to 2. Why not? Because 1 has order 4 and 2 has order 2. 2 is the only element that is order 2, and so 2 must be mapped to 2. Hence we are left with only two possibilities for the image of 1: either 1 itself or 3. Now if 1 is mapped to 1, then $\phi(3) = \phi(1 + 2) = \phi(1) + \phi(2) = 1 + 2 = 3$, and this is the identity map. On the other hand, if 1 goes to 3, then $\phi(3) = \phi(1 + 2) = \phi(1) + \phi(2) = 3 + 2 = 1$. This time the automorphism fixes 0 and 2 and simply interchanges 1 and 3: $\phi = (1, 3)$ in cycle notation. Again, there are only two possible automorphisms of Z_4.

We will consider one more easy example, the Klein 4-group **V**. This time we will name the elements 1, a, b, and c using multiplicative notation. Any automorphism must map 1 to 1, but there is considerable freedom in mapping a, b and c because they are all order 2. Now **V** is not cyclic so it requires two generators – in fact, any two of a, b or c can act as generators and we will choose a and b to be the generators now. Begin with $\phi(a) = a$; then we still have a choice for b. If $\phi(b) = b$ then we are forced to map c to c and we get the identity map. If $\phi(b) = c$ then $\phi(c) = \phi(ab) = \phi(a)\phi(b) = ac = b$. In this case we have determined ϕ = (b, c) with 1 and a both fixed. There are going to be other automorphisms so we will call this automorphism ϕ_1.

Now suppose ϕ_2(a) = b. We still have two choices of where to map b. First suppose ϕ_2(b) = c and consider what ϕ_2(c) must equal. To be a bijection, ϕ_2(c) must equal a but we will compute it algebraically just to make sure it preserves the algebra (this will not be necessary for much longer): ϕ_2(c) = ϕ_2(ab) = ϕ_2(a)ϕ_2(b) = bc = a. In cycle notation this is ϕ_2 = (a, b, c), a distinct automorphism from ϕ_1. Now try the other choice for b.

Still supposing ϕ_3(a) = b, let ϕ_3(b) = a. Then we must assign ϕ_3(c) = c and this time I will leave checking the products to you. We get ϕ_3 = (a, b).

The last case to consider is to suppose ϕ_4(a) = c. Then we know b can be mapped to either a or itself. The computations involved should be getting a bit routine by now and I will leave it to you to fill in that we get two more automorphisms: ϕ_4 = (a, c, b) and ϕ_5 = (a, c).

Perhaps you are noticing there is a pattern in what we have been calculating. It is summarized in the following:

Theorem 46: Let G be a group. Then the set of all automorphisms of G form a group under composition, denoted by **Aut(G).**

Proof: First, the identity map has already been singled out as an automorphism.

Second, the composition of two isomorphisms is an isomorphism, and therefore the composition of two automorphisms is also an automorphisms.

Third, the inverse of an isomorphism is an isomorphism, so the inverse of an automorphism is an automorphism.

In the examples above, we have shown the following three facts: **Aut(Z_3)** ≃ Z_2, **Aut(Z_4)** ≃ Z_2 and **Aut(V)** ≃ D_3. Since automorphisms are permutations, if $|G|$ = n then we know **Aut(G)** ≤ S_n. In fact, because automorphisms fix the identity element, we can be more precise: **Aut(G)** ≤ $Stab$(1) ≃ S_{n-1}. We will be interested in computing **Aut**(G) for a given G, but this is a very difficult problem in general. In the next lesson we will look at the easiest case, the automorphism group of cyclic groups.

That **Aut**(G) is a group is one of the secrets to the fruitfulness of group theory. We began by defining an algebraic structure on a set. Then we considered particular functions on that set that respected the algebraic structure and discovered, enchantingly, that the collection of functions shared the same algebraic structure as the sets they mapped. This is a remarkable pattern that recurs constantly in algebra, and frequently in other mathematical disciplines as well.

Exercise 23: Find **Aut(Z_5)** and **Aut(Z_6)**.
Exercise 24: Find **Aut(Z_8)** and **Aut(Z_9)**.

Lesson 8: Aut(Z_n)

An automorphism is completely determined by how it maps the generators of the group. Since cyclic groups have only one generator, the automorphism group of a cyclic group has a particularly easy structure to determine. In this lesson we will consider in detail the automorphism group of Z_{12} to get a glimpse of how an automorphism may rearrange the internal structure of a group and simultaneously preserve all the algebra.

We will take Z_{12} in additive notation, generated by 1. Any automorphism must map 1 to another element of order 12; that is, any automorphism must map 1 to another generator of Z_{12}. We know that the generators of a cyclic group Z_n in additive notation are the numbers that are relatively prime to n. In this case, the numbers relatively prime to 12 are - besides 1 – the numbers 5, 7, and 11. Thus we expect exactly three non-trivial automorphisms of Z_{12}. If $\phi(1) = 1$, we get the identity automorphism; we will use a subscript and denote this automorphism by ϕ_0. The non-trivial automorphisms are determined by $\phi_1(1) = 5$, $\phi_2(1) = 7$, and $\phi_3(1) = 11$. Here are the details of how each of these automorphisms is determined completely by choosing how it maps the generator:

$\phi_1(1) = 5$
$\phi_1(2) = \phi_1(1 + 1) = \phi_1(1) + \phi_1(1) = 5 + 5 \equiv 10 \pmod{12}$
$\phi_1(3) = \phi_1(2 + 1) = \phi_1(2) + \phi_1(1) = 10 + 5 \equiv 3 \pmod{12}$
$\phi_1(4) = \phi_1(3 + 1) = \phi_1(3) + \phi_1(1) = 3 + 5 \equiv 8 \pmod{12}$
$\phi_1(5) = \phi_1(4 + 1) = \phi_1(4) + \phi_1(1) = 8 + 5 \equiv 1 \pmod{12}$
$\phi_1(6) = \phi_1(5 + 1) = \phi_1(5) + \phi_1(1) = 1 + 5 \equiv 6 \pmod{12}$
$\phi_1(7) = \phi_1(6 + 1) = \phi_1(6) + \phi_1(1) = 6 + 5 \equiv 11 \pmod{12}$
$\phi_1(8) = \phi_1(7 + 1) = \phi_1(7) + \phi_1(1) = 11 + 5 \equiv 4 \pmod{12}$
$\phi_1(9) = \phi_1(8 + 1) = \phi_1(8) + \phi_1(1) = 4 + 5 \equiv 9 \pmod{12}$
$\phi_1(10) = \phi_1(9 + 1) = \phi_1(9) + \phi_1(1) = 9 + 5 \equiv 2 \pmod{12}$
$\phi_1(11) = \phi_1(10 + 1) = \phi_1(10) + \phi_1(1) = 2 + 5 \equiv 7 \pmod{12}$
$\phi_1(12) = \phi_1(0) = \phi_1(11 + 1) = \phi_1(11) + \phi_1(1) = 7 + 5 \equiv 0 \pmod{12}$ (as required)

In less detail, the other two automorphisms are given by the following:

$\phi_2(1) = 7$
$\phi_2(2) = 7 + 7 \equiv 2 \pmod{12}$
$\phi_2(3) = 2 + 7 \equiv 9 \pmod{12}$
$\phi_2(4) = 9 + 7 \equiv 4 \pmod{12}$
$\phi_2(5) = 4 + 7 \equiv 11 \pmod{12}$
$\phi_2(6) = 11 + 7 \equiv 6 \pmod{12}$
$\phi_2(7) = 6 + 7 \equiv 1 \pmod{12}$
$\phi_2(8) = 1 + 7 \equiv 8 \pmod{12}$
$\phi_2(9) = 8 + 7 \equiv 3 \pmod{12}$
$\phi_2(10) = 3 + 7 \equiv 10 \pmod{12}$
$\phi_2(11) = 10 + 7 \equiv 5 \pmod{12}$
$\phi_2(0) = 5 + 7 \equiv 0 \pmod{12}$

$\phi_3(1) = 11$
$\phi_3(2) = 11 + 11 \equiv 10 \pmod{12}$
$\phi_3(3) = 10 + 11 \equiv 9 \pmod{12}$
$\phi_3(4) = 9 + 11 \equiv 8 \pmod{12}$
$\phi_3(5) = 8 + 11 \equiv 7 \pmod{12}$
$\phi_3(6) = 7 + 11 \equiv 6 \pmod{12}$
$\phi_3(7) = 6 + 11 \equiv 5 \pmod{12}$
$\phi_3(8) = 5 + 11 \equiv 4 \pmod{12}$
$\phi_3(9) = 4 + 11 \equiv 3 \pmod{12}$
$\phi_3(10) = 3 + 11 \equiv 2 \pmod{12}$
$\phi_3(11) = 2 + 11 \equiv 1 \pmod{12}$
$\phi_3(0) = 1 + 11 \equiv 0 \pmod{12}$

What is of interest here is how an automorphism may rearrange the elements within a particular subgroup without moving the elements out of that subgroup. An automorphism may leave all the elements of a particular subgroup fixed and rearrange the elements of another subgroup. In the diagrams below, the subgroups whose elements are rearranged are indicated by the arrows circling back in to it, and the subgroups which are left entirely fixed have no such arrows. Some contemplation of the diagrams may help you understand in more detail how automorphisms work.

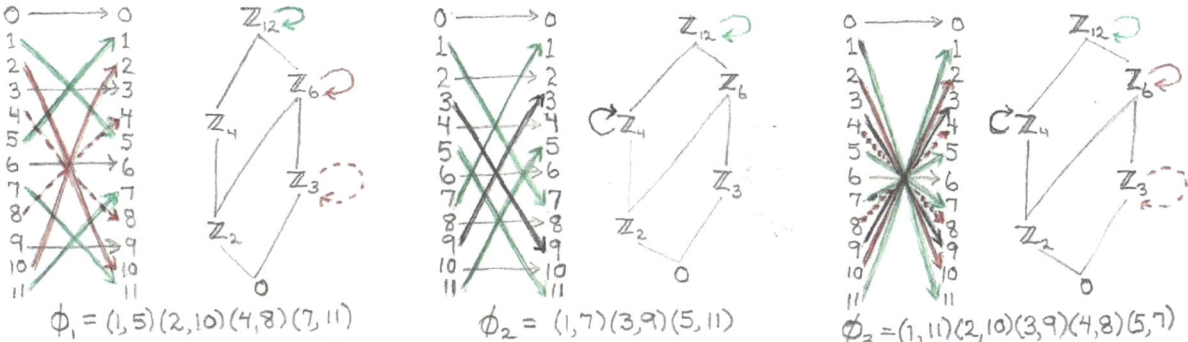

$\phi_1 = (1,5)(2,10)(4,8)(7,11)$ $\phi_2 = (1,7)(3,9)(5,11)$ $\phi_3 = (1,11)(2,10)(3,9)(4,8)(5,7)$

Not only are the automorphism groups of cyclic groups fairly easy to calculate, they are so easy that we can prove a general theorem about them.

Theorem 47: $\mathbf{Aut}(\mathbf{Z_n}) \simeq \mathbf{U_n}$.

Proof: We will prove this by defining the isomorphism explicitly. First notice that the two groups in question are the same size. There is one automorphism for each element of $\mathbf{Z_n}$ which is relatively prime to n; that is, there are $\varphi(n)$ automorphisms. Also $\mathbf{U_n}$ consists of those elements of $\mathbf{Z_n}$ which are units, and these are exactly the numbers that are relatively prime to n; again, there are $\varphi(n)$ of them.

Let k, m $\in \mathbf{U_n}$ and let ϕ_k be the automorphism of $\mathbf{Z_n}$ determined by $\phi_k(1) = k$; and ϕ_m the automorphism determined by $\phi_m(1) = m$. Now define an isomorphism $\psi: \mathbf{Aut}(\mathbf{Z_n}) \to \mathbf{U_n}$ in the most straight-forward way, by $\psi(\phi_k) = k$. ψ is plainly a bijection. All we need to show is that ψ preserves products; that is, $\psi(\phi_k \circ \phi_m) = k \cdot m \pmod{n}$. We need to compute $\phi_k \circ \phi_m(1)$ in order to find its image under ψ.

$\phi_k \circ \phi_m(1) = \phi_k(\phi_m(1)) = \phi_k(m) = \underbrace{\phi_k(1+1+\ldots+1)}_{m \text{ times}} = \underbrace{\phi_k(1) + \phi_k(1) + \ldots + \phi_k(1)}_{m \text{ times}}$

$= \underbrace{k + k + \ldots + k}_{m \text{ times}} = k \cdot m \pmod{n} = \phi_{km}(1)$

Therefore $\phi_k \circ \phi_m = \phi_{km}$ and we have shown that $\psi(\phi_k \circ \phi_m) = \psi(\phi_{km}) = km = \psi(\phi_k) \cdot \psi(\phi_m)$. Thus ψ is an isomorphism.

Unfortunately, cyclic groups are some of the very few groups for which the automorphism group is easy to compute. There is no general method at the moment for calculating $\mathbf{Aut}(G)$ in general. In the next lesson we will consider some of the things that may happen with an automorphism of a non-cyclic or non-Abelian group.

Exercise 25: Find $\mathbf{Aut}(\mathbf{Z_{777}})$. Put the answer in standard form for direct products of cyclic groups.

Exercise 26: Find $\mathbf{Aut}(\mathbf{Z_{24650}})$. Put the answer in standard form for direct products of cyclic groups.

Lesson 9: More Complicated Automorphism Groups

We will do one more lesson on automorphisms, this time with some non-Abelian groups, to get a more complete view of how automorphisms behave. Unfortunately, there is no general easy method for discovering the automorphism group of a non-cyclic group, much less a non-Abelian group. In this lesson we will compute **Aut(Q_8)** and **Aut(D_4)**.

To compute the automorphism group of any group we should first choose a presentation because everything hinges on how we choose to map the generators. For the quaternions we will use this one: $Q_8 = <a, b : a^4 = b^4 = 1, a^2 = b^2, ab = (ba)^{-1}>$. We will also use $c = ab$ for convience. We determine an automorphism by assigning a value to each generator, so we will proceed in a systematic way, first assigning a and then considering each possible assignment for b. Since we know a, b, and ab are all order 4, there are six possible elements to assign to a: $\phi_1(a) = a$, $\phi_2(a) = a^3$, $\phi_3(a) = b$, $\phi_4(a) = b^3$, $\phi_5(a) = c$, and $\phi_6(a) = c^3$.

Case 1: Let $\phi_1(a) = a$. Can $\phi_1(b) = a^3$? No, because $\phi_1(a^3)$ must equal a^3, and since ϕ_1 is injective that would require $b = a^3$, which is impossible. So there are only 4 possible images for b: $\phi_{1,1}(b) = b$, $\phi_{1,2}(b) = b^3$, $\phi_{1,3}(b) = c$ and $\phi_{1,4}(b) = c^3$.

The first possibility will produce the identity automorphism. Next, consider $\phi_{1,2}(b) = b^3$; then $\phi_{1,2}(b^3) = b$, $\phi_{1,2}(c) = \phi_{1,2}(ab) = ab^3 = c^3$, and finally $\phi_{1,2}(c^3) = c$. Thus $\phi_{1,2} = (b, b^3)(c, c^3)$ in cycle notation.

If we consider $\phi_{1,3}(b) = c$, then we know $\phi_{1,3}(b^3) = c^3$. We then calculate $\phi_{1,3}(c) = \phi_{1,3}(ab) = ac = b^3$, and this requires $\phi_{1,3}(c^3) = b$. This gives $\phi_{1,3} = (b, c, b^3, c^3)$ in cycle notation.

Finally $\phi_{1,4}(b) = c^3$, by the same steps, is seen to be $\phi_{1,4} = (b, c^3, b^3, c)$.

Case 2: Next let $\phi_2(a) = a^3$. Then immediately we know $\phi_2(a^3) = a$. We have four possibilities for the other generator: $\phi_{2,1}(b) = b$, $\phi_{2,2}(b) = b^3$, $\phi_{2,3}(b) = c$, $\phi_{2,4}(b) = c^3$.

If $\phi_{2,1}(b) = b$ then automatically $\phi_{2,1}(b^3) = b^3$, and $\phi_{2,1}(c) = \phi_{2,1}(ab) = a^3b = c^3$, and so $\phi_{2,1}(c^3) = c$. Hence $\phi_{2,1} = (a, a^3)(c, c^3)$.

If $\phi_{2,2}(b) = b^3$ then also automatically $\phi_{2,2}(b^3) = b$; $\phi_{2,2}(c) = \phi_{2,2}(ab) = a^3b^3 = c$ and finally $\phi_{2,2}(c^3) = c^3$. Thus $\phi_{2,2} = (a, a^3)(b, b^3)$.

Next, if $\phi_{2,3}(b) = c$, then $\phi_{1,3}(b^3) = c^3$. So $\phi_{2,3}(c) = \phi_{2,3}(ab) = a^3c = b$ and then $\phi_{2,3}(c^3) = b^3$. This gives $\phi_{2,3} = (a, a^3)(b, c)(b^3, c^3)$.

The same sequence of steps gives $\phi_{2,4} = (a, a^3)(b, c^3)(b^3, c)$.

Cases 3 and 4: There are many more calculations to go, but if you have followed all the steps above, the rest should be routine. I will list only the results for the remaining automorphisms in cycle notation and leave the work as an exercise:

$\phi_{3,1} = (a, b)(a^3, b^3)(c, c^3)$ $\phi_{3,2} = (a, b, a^3, b^3)$ $\phi_{3,3} = (a, b, c)(a^3, b^3, c^3)$ $\phi_{3,4} = (a, b, c^3)(a^3, b^3, c)$
$\phi_{4,1} = (a, b^3, a^3, b)$ $\phi_{4,2} = (a, b^3)(a^3, b)(c, c^3)$ $\phi_{4,3} = (a, b^3, c^3)(a^3, b, c)$ $\phi_{4,4} = (a, b^3, c)(a^3, b, c^3)$
$\phi_{5,1} = (a, c, b)(a^3, c^3, b^3)$ $\phi_{5,2} = (a, c, b^3)(a^3, c^3, b)$ $\phi_{5,3} = (a, c, a^3, c^3)$ $\phi_{5,4} = (a, c)(a^3, c^3)(b, b^3)$
$\phi_{6,1} = (a, c^3, b)(a^3, c, b^3)$ $\phi_{6,2} = (a, c^3, b^3)(a^3, c, b)$ $\phi_{6,3} = (a, c^3, a^3, c)$ $\phi_{6,4} = (a, c^3)(a^3, c)(b, b^3)$

Thus, we began with a group of order 8 and found 24 automorphisms. The automorphism group is quite a lot more complex than the original group, and this is not rare. As a long exercise, you will be asked to find the lattice diagram of this automorphism group. Sorting the elements into cyclic groups is not difficult, but then you must look for all the copies of **V** that are present and also the dihedral subgroups. If you find all of these you will be doing well. There are other subgroups of **Aut(Q_8)**, but it is a long enough exercise just to find these. On the next page are the diagrams, in lattice form, of these 24 automorphism. It will be of some help in looking for the Klein 4-subgroups, and perhaps the dihedral subgroups as well.

As one more exercise to do with **Aut(Q₈)**, consider the symmetries of the cube. In lesson 27 of part 1, we found that the group **O*₄₈** of symmetries of the cube has order 48. But it has a subgroup of order 24 which consists in the identity element plus all the rotations. This subgroup is called the *group of rigid motions* of the cube. In the exercises you will be asked to attempt to determine whether or not the group of rigid motions of the cube is isomorphic to the group of automorphisms of the quaternions? At the least you should be able to define a candidate isomorphism.

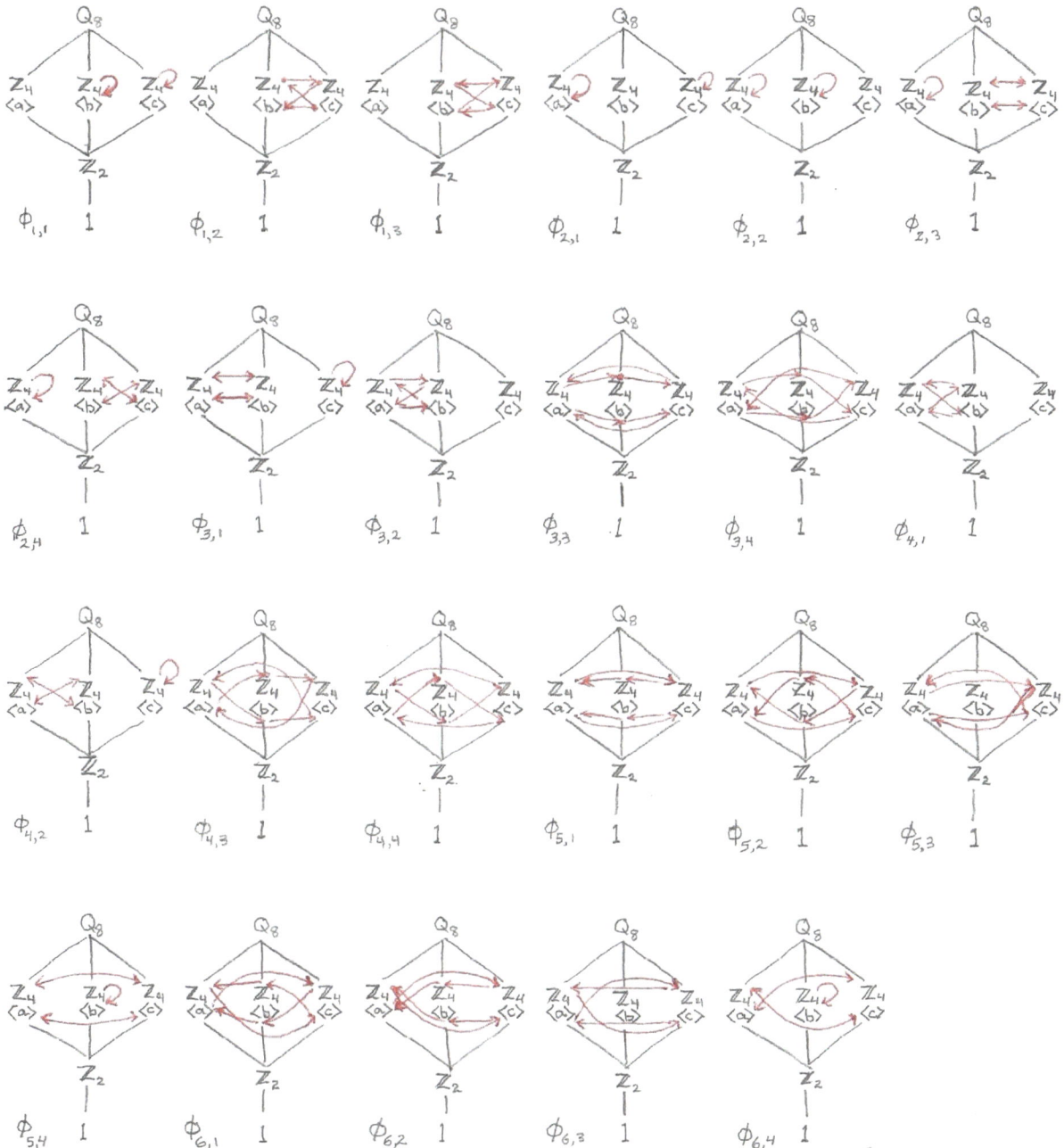

Though **D₄** has the same order as **Q₈**, **Aut(D₄)** is considerably easier to calculate. First, choose a presentation for **D₄**. Using the common names for the elements, the most convenient presentation for the present purpose is $\mathbf{D_4} = <\mu_1, \delta_1 : \mu_1^2 = \delta_1^2 = 1, (\mu_1 \cdot \delta_1)^4 = 1>$. We will look at the possible images of μ_1 and try each one with the possible images of δ_1. The choices for the two generators will then determine $\mu_1 \cdot \delta_1$, which can be either element of order 4, the only difference being whether we have chosen a clockwise or counter-clockwise rotation. Note that the choice for the order 2 generators had to be a μ and a δ but it didn't matter which pair we chose to begin with.

Before trying the possible ways of mapping these order 2 generators to other order 2 elements we can save ourselves some work by noting that neither can map to ρ_2. ρ_2 is the 180° rotation and forms a subgroup that is contained in both copies of **V** and the copy of $\mathbf{Z_4}$. That structure must be preserved by any automorphism. Mapping a reflection to ρ_2 would destroy the intersection of the three subgroups. Further, if μ_1 maps to a δ then the other μ must also map to a δ in order to preserve their common subgroup **V**. So we have these restrictions on possible automorphisms: either both μ's map to μ's (and therefore both δ's map to δ's), or the μ's and the δ's are interchanged. This gives us the following possibilities:

$\phi_0(\mu_1) = \mu_1$ and $\phi_0(\delta_1) = \delta_1$ (which is the identity automorphism)

$\phi_1(\mu_1) = \mu_1$ and $\phi_1(\delta_1) = \delta_2 \Rightarrow \phi_1(\rho_3) = \phi_1(\mu_1 \cdot \delta_1) = \phi_1(\mu_1) \cdot \phi_1(\delta_1) = \mu_1 \cdot \delta_2 = \rho_1$, with $\phi_1(\rho_1) = \rho_3$

$\phi_2(\mu_1) = \mu_2$ and $\phi_2(\delta_1) = \delta_1 \Rightarrow \phi_2(\rho_3) = \phi_2(\mu_1 \cdot \delta_1) = \phi_2(\mu_1) \cdot \phi_2(\delta_1) = \mu_2 \cdot \delta_1 = \rho_1$, with $\phi_2(\rho_1) = \rho_3$

$\phi_3(\mu_1) = \mu_2$ and $\phi_3(\delta_1) = \delta_2 \Rightarrow \phi_3(\rho_3) = \phi_3(\mu_1 \cdot \delta_1) = \phi_3(\mu_1) \cdot \phi_3(\delta_1) = \mu_2 \cdot \delta_2 = \rho_3$, with $\phi_3(\rho_1) = \rho_1$

$\phi_4(\mu_1) = \delta_1$ and $\phi_4(\delta_1) = \mu_1 \Rightarrow \phi_4(\rho_3) = \phi_4(\mu_1 \cdot \delta_1) = \phi_4(\mu_1) \cdot \phi_4(\delta_1) = \delta_1 \cdot \mu_1 = \rho_1$, with $\phi_4(\rho_1) = \rho_3$

$\phi_5(\mu_1) = \delta_1$ and $\phi_5(\delta_1) = \mu_2 \Rightarrow \phi_5(\rho_3) = \phi_5(\mu_1 \cdot \delta_1) = \phi_5(\mu_1) \cdot \phi_5(\delta_1) = \delta_1 \cdot \mu_2 = \rho_3$, with $\phi_5(\rho_1) = \rho_1$

$\phi_6(\mu_1) = \delta_2$ and $\phi_6(\delta_1) = \mu_1 \Rightarrow \phi_6(\rho_3) = \phi_6(\mu_1 \cdot \delta_1) = \phi_6(\mu_1) \cdot \phi_6(\delta_1) = \delta_2 \cdot \mu_1 = \rho_3$, with $\phi_6(\rho_1) = \rho_1$

$\phi_7(\mu_1) = \delta_2$ and $\phi_7(\delta_1) = \mu_2 \Rightarrow \phi_7(\rho_3) = \phi_7(\mu_1 \cdot \delta_1) = \phi_7(\mu_1) \cdot \phi_7(\delta_1) = \delta_2 \cdot \mu_2 = \rho_1$, with $\phi_7(\rho_1) = \rho_3$

The lattice diagrams for these automorphisms are on the next page.

When these automorphisms are written in cycle notation, we see something interesting. Examining these cycles, you should be immediately suspect that $\mathbf{Aut(D_4)} \simeq \mathbf{D_4}$. Thus we have groups whose automorphisms form a smaller group, groups whose automorphisms form a much larger and more complicated group, and groups whose automorphisms form the same group.

One way to think of **Aut**(G) is as the group of symmetries of the group G. As with polygons and other geometric shapes, the group of symmetries of those shapes preserved their geometric character. It is the similar here; the group of symmetries of an algebraic structure like a group preserve their algebraic character and their geometric structure in so far as that is revealed in the lattice diagram. In the case of $\mathbf{D_4}$, which is already a group of symmetries of the square, $\mathbf{Aut(D_4)}$ is the group of "second order" symmetries of the square – the symmetries of the symmetries. Before long we will see that any group can be understood as the group of symmetries of some object, and the automorphism group can always be understood as the group of second order symmetries.

Exercise 27: Show that $\mathbf{Aut(D_4)} \simeq \mathbf{D_4}$.

Exercise 28: Find $\mathbf{Aut(D_5)}$.

Exercise 29: Find $\mathbf{Aut(D_6)}$.

Exercise 30: Find the structure of $\mathbf{Aut(Q_8)}$, as described in the text above and try to show it is isomorphic to the group of rigid motions of the cube.

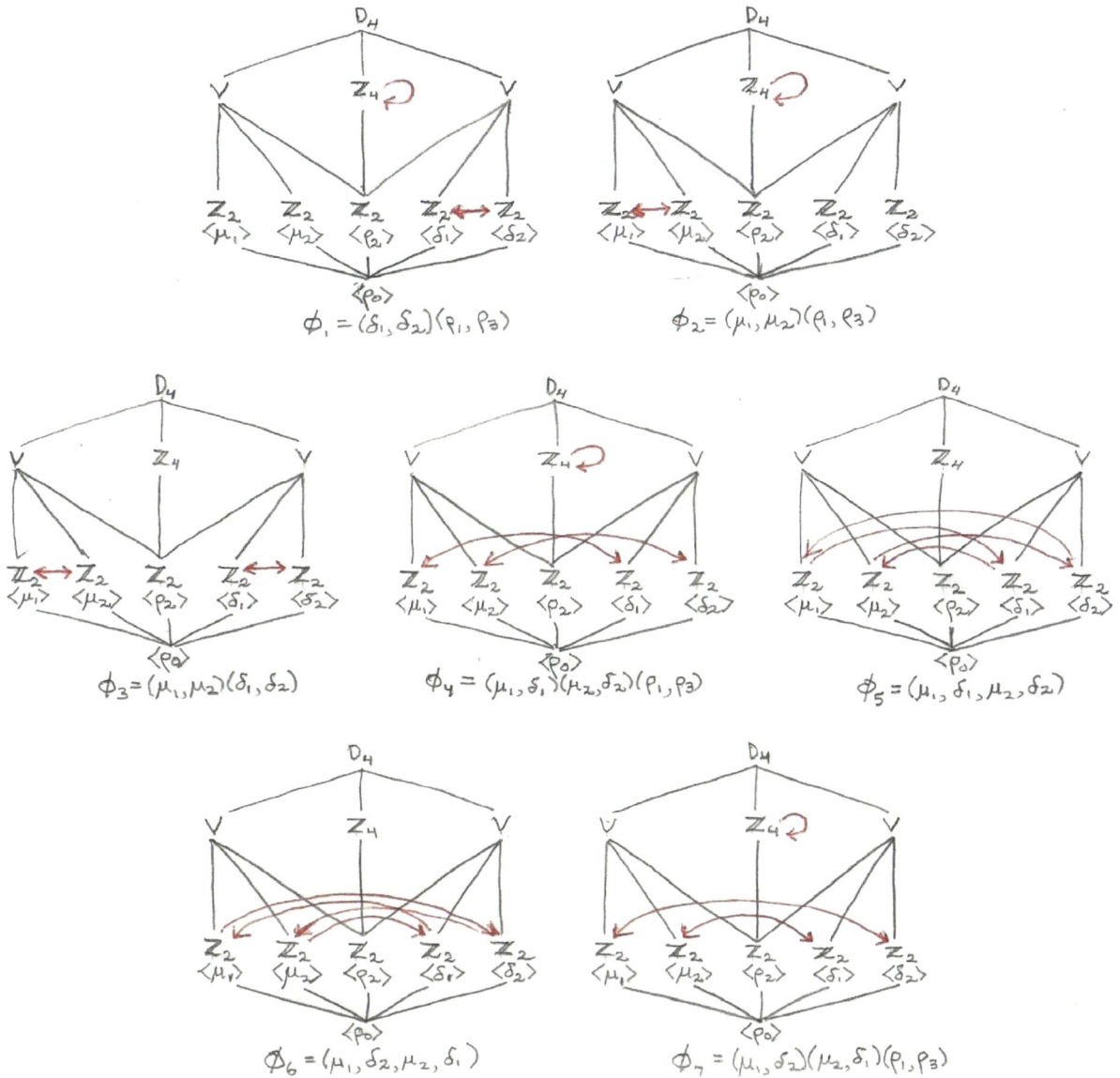

Lesson 10: Characteristic Subgroups

Suppose H ≤ G and suppose that G has no other subgroup which is isomorphic to H. If β is an automorphism of G, then we must have β(H) = H. Since cyclic groups Z_n have only one subgroup for each order dividing n, any automorphism of Z_n must necessarily map each subgroup onto itself; the automorphisms of cyclic groups permute the elements of each of the subgroups only within their subgroups. This is not the case with non-cyclic groups like **V**, or D_4, or Q_8. Subgroups may be interchanged with other subgroups isomorphic to them under certain conditions.

But there are restrictions that may not be immediately visible and that prevent some subgroup or other from being mapped to any other subgroup like it. D_4 is a bit more complicated. There is only one Z_4, the subgroup of rotations, and that must be mapped into itself, but there were two copies of **V** which then could potentially be exchanged. Those two copies of **V**, however, intersected each other in a single Z_2, the subgroup generated by the 180° rotation. In order to preserve both copies of **V**, the intersection of the two had to be preserved. That particular copy of Z_2 could not be mapped to any other copy of Z_2 without violating the structure of both copies of **V**. The other four copies of Z_2 could be shuffled around respectively, only making sure that neither copy of **V** got split up. We could not, for example, map a μ reflection to a δ reflection and then map the other μ reflection to a μ reflection; that would shred the **V** containing the μ's. Only the Z_2 containing the 180° rotation had to be fixed by every automorphism.

In general, automorphisms will sometimes fix a particular subgroup by mapping it only into itself; and sometimes an automorphism may interchange two identical subgroups without violating the structure. Consider this analogy with architectural structures. Every house has many rooms, but some of the rooms are uniquely suited to one use only. A bathroom is a bathroom and cannot be a dining room. A kitchen is a kitchen and not a bedroom. A hallway must remain a hallway. But the remaining rooms of the house can be assigned their role somewhat arbitrarily. In a similar way, some subgroups play an essential and invariable role in the algebra, and some have a bit of flexibility built into them. Here we are interested in those subgroups which play the most essential and invariable roles in the group structure.

Definition 33: Let G be a group and H ≤ G. When $β(H) = H$ $∀β ∈$ **Aut**(G), then we call H a **characteristic subgroup** of G, and we denote it by writing H ≪ G.

The notation I am using for characteristic subgroups is not standard; I think it is more convenient however. The most common notation is to write "H char G", which is just unfortunate.

A characteristic subgroup of G is a subgroup which is *always* preserved by *every* automorphism of the group. Characteristic subgroups play an important role in the structure of a group. In the remainder of this part of <u>Modern Algebra</u> we will have occasion to study four characteristic subgroups that determine many properties and a large part of the structure of any group. At this point we need to learn the fundamental properties that all characteristic subgroups share.

First we know that *for cyclic groups, every subgroup is characteristic*. This is a rare property among groups. Eventually we will determine all the groups which have this property. It is mainly important, at this point, to note that Abelian groups in general do not share this property. For example, the Klein 4-group has no proper characteristic subgroup. The trivial subgroup, of course, is always characteristic; the improper subgroup is always characteristic; it is the proper characteristic subgroups that are interesting.

To discuss characteristic subgroups further, we will need some terminology which you should have seen before but which I will repeat for review.

Definition 34: Let f: X → Y be any function defined on the sets X and Y, and let A ⊆ X. The function $f|_A$: A → Y defined by $f|_A(x) = f(x)$ ∀x ∈ A is called the **restriction of f to A**.

f|_A is essentially the same function as f but with a smaller domain. The next theorem is easy to prove so its proof will be left as an exercise.

Theorem 48: Let G be a group, $H \ll G$, and $\beta \in \mathbf{Aut}(G)$. Then $\beta|_H \in \mathbf{Aut}(H)$.
Proof: is an exercise.

Note that if H is not characteristic, then there will be automorphisms which map H to some other subgroup and so the restriction of those automorphisms to H will also map H to some other subgroup. But the restriction of an automorphism to a characteristic subgroup is still an automorphism, but an automorphism of that subgroup. This enables us to easily prove another important property of characteristic subgroups, which is also left as an exercise.

Theorem 49: Let G be a group, $H \ll G$, and $K \ll H$. Then $K \ll G$.
Proof: is an exercise.

In words, the characteristic subgroup of a characteristic subgroup is a characteristic subgroup. Being characteristic is a transitive relation. This is important. When transitivity fails to hold for a relation, things get much more complicated (and more interesting), as we shall soon see.

To relate this new idea to what we have learned already, the question is: are the factors of a direct product always characteristic subgroups or not? This question is one that you can answer in the exercises. We do get one more result connecting characteristic subgroups and direct products.

Theorem 50: Let G be any group and let H and K be two characteristic subgroups such that $H \cap K = \mathbf{1}$ and such that $H \vee K = G$. Then $\mathbf{Aut}(G) \simeq \mathbf{Aut}(H) \times \mathbf{Aut}(K)$.

Proof: We will prove this, as usual, by giving the isomorphism explicitly. Before we can do so, we need to provide ourselves with a way of factoring automorphisms of G. Let β be any automorphism of G and define two other functions from β. First define $\beta'(x) = \beta(x) \,\forall x \in H$ and $\beta'(x) = x \,\forall x \in K$; similarly, define $\beta''(x) = \beta(x) \,\forall x \in K$ and $\beta'(x) = x \,\forall x \in H$. β' agrees with β on H but is the identity on K, while β'' agrees with β on K but is the identity on H. I claim that both β' and β'' are also automorphisms of G. Since H and K are both characteristic, $\beta'(H) = H$ and $\beta''(K) = K$. β' and β'' are both clearly bijections.

Since $H \vee K = G$, each element in G can be expressed as a product of elements of H with elements of K, so the definitions of both β' and β'' are defined over all of G; they are well-defined because $H \cap K = \mathbf{1}$ and so there will never be ambiguity when we extend them over all of G. They will each clearly preserve products because of the way they are defined, and so they are each automorphisms themselves.

Further, for any automorphism $\beta \in \mathbf{Aut}(G)$, $\beta = \beta' \circ \beta'' = \beta'' \circ \beta'$. We can easily check this by choosing an arbitrary element of G, which has the form $h_1 k_1 h_2 k_2 \ldots h_n k_n$ and calculating

$(\beta'' \circ \beta')(h_1 k_1 h_2 k_2 \ldots h_n k_n) = \beta''(\beta'(h_1 k_1 h_2 k_2 \ldots h_n k_n))$
$$= \beta''(\beta'(h_1)\beta'(k_1)\beta'(h_2)\beta'(k_2)\ldots\beta'(h_n)\beta'(k_n))$$
$$= \beta''(\beta(h_1)k_1\beta(h_2)k_2\ldots\beta(h_n)k_n)$$
$$= \beta''(\beta(h_1))\beta''(k_1)\beta''(\beta(h_2))\beta''(k_2)\ldots\beta''(\beta(h_n))\beta''(k_n)$$
$$= \beta(h_1)\beta(k_1)\beta(h_2)\beta(k_2)\ldots\beta(h_n)\beta(k_n)$$
$$= \beta(h_1 k_1 h_2 k_2 \ldots h_n k_n)$$

Since β' and β'' are both automorphisms on G, theorem 48 tells us that $\beta'|_H$ is an automorphism of H and $\beta''|_K$ is an automorphism of K. We can now define a candidate isomorphism
$$\Phi: \mathbf{Aut}(G) \to \mathbf{Aut}(H) \times \mathbf{Aut}(K) \text{ by } \Phi(\beta) = (\beta', \beta'')$$

By construction, Φ is a bijection. We need only verify that it preserves products. To do this we must first note that $(\beta \circ \gamma)'$ agrees with $\beta \circ \gamma$ on H and is the identity on K. Further
$$(\beta' \circ \gamma')(h) = \beta'(\gamma'(h)) = \beta'(\gamma(h)) = \beta(\gamma(h))$$

Thus β'∘γ' agrees with β∘γ on H. β'∘γ' is clearly the identity on K, and so β'∘γ' = (β∘γ)'. Similarly (β∘γ)" = β"∘γ". Now we are in the position to calculate:
$$\Phi(\beta\circ\gamma) = ((\beta\circ\gamma)', (\beta\circ\gamma)'') = ((\beta'\circ\gamma'), (\beta''\circ\gamma'')) = (\beta', \beta'')(\gamma', \gamma'') = \Phi(\beta)\Phi(\gamma)$$
Hence Φ is an isomorphism and we are done.

Since the factor groups of a direct product are not necessarily characteristic, theorem 50 does not, unfortunately, give us a way of computing the automorphism groups of all Abelian groups, as you will be asked to show in the exercises.

Exercise 31: Prove theorem 48.

Exercise 32: Prove theorem 49.

Exercise 33: Let $G \simeq H \times K$. Is it true that $H \ll G$? Prove that it is or else give a counter-example.

Exercise 34: Find an Abelian group $H \times K$ for which **Aut**$(H \times K)$ and **Aut**$(H) \times$ **Aut**(K) are not isomorphic.

Lesson 11: Cosets

The best way to continue analyzing the subgroup structure is, oddly enough, to back off a bit and introduce two functions that are *not* automorphisms. These are the simplest possible non-trivial functions that can be defined on a group. Let G be a group, and let g ∈ G. First we define the *left multiplication by g* as the function $\lambda_g : G \to G$ given by $\lambda_g(x) = gx$ $\forall x \in G$. Second we define the *right multiplication by g* as the function $\rho_g : G \to G$ given by $\rho_g(x) = xg$ $\forall x \in G$.

Now it is clear that the cancellation laws guarantee both of these functions are bijections. It is also clear that neither of these functions is an automorphism because neither will preserve products:
$\lambda_g(xy) = gxy$ (by the definition of λ_g)
$\lambda_g(x) \cdot \lambda_g(y) = (gx) \cdot (gy) = gxgy$ or g^2xy (depending on whether G is Abelian or not)

As simple as the left and right multiplications are, and though they do not preserve the algebra within the group, they do lead to remarkable and unexpected insight into the structure. In fact, it is the insights gained from these two functions that will occupy us through the rest of this book. We will not so much be interested in how these functions map particular elements of the group – that is just the group operation, after all. Instead we will focus on how these functions map whole subgroups. Not preserving the algebra, we do not expect them to map a subgroup to another subgroup, and what value a random set of elements may have we may not yet foresee; nonetheless we persist.

Definition 35: Let G be a group and H ≤ G. $\lambda_g(H) = \{gh \mid h \in H\}$ is called the **left coset of H by g** and $\rho_g(H) = \{hg \mid h \in H\}$ is called the **right coset of H by g**. The left coset of H by g is denoted as gH, and the right coset of H by g is denoted as Hg.

If we use additive notation, a left coset would be written g+H and a right coset would be written H+g. From this point on we will be more strict about to using capital letters for subgroups and subsets, and lower case letters for elements. It is important to get firmly in mind that gH and Hg are entire subsets of G, and not single elements. Here are the simplest properties of cosets in a single theorem.

Theorem 51: Let G be a group and H ≤ G. Then
    a) either gH = kH or gH ∩ kH = ø.
    b) $|H| = |gH| = |Hg|$ $\forall g \in G$
    c) gH = H iff g ∈ H. Similarly Hg = H iff g ∈ H.
    d) gH = kH iff g = kh for some h ∈ H.

Proof: a) let gH ∩ kH ≠ ø. Then there is some element x that is both in gH and in kH. Now if x ∈ gH then there is some $h_1 \in H$ such that $\lambda_g(h_1) = x$, or by definition 34 $gh_1 = x$. In the same way x ∈ kH means there exists some $h_2 \in H$ such that $\lambda_k(h_2) = x$, or $kh_2 = x$. Therefore we have $gh_1 = kh_2$. Hence $g = kh_2h_1^{-1}$. Since $h_2h_1^{-1} \in H$, this means g ∈ kH. Now let h' be an arbitrary element of H, so that gh' is an arbitrary element of gH. Then gh' = $kh_2h_1^{-1}$h' and we see that gh' ∈ kH as well. Since gh' is an arbitrary element of gH, every element of gH is included in kH and gH ⊆ kH. We can repeat the same argument beginning from $gh_1 = kh_2$, solving for k instead of for g, and conclude that kH ⊆ gH. Therefore gH = kH and part a) is proved.

The proofs of parts **b)**, **c)** and **d)** are left as exercises.

Theorem 51a can be restated this way:
    the collection of left (or right) cosets of the subgroup H is a *partition* of G.

In fact, according to 49b, this is a partition into equal sized subsets. Theorem 51d can be restated in several equivalent ways, any of which may be useful. Consider the following restatements:
    gH = kH iff g ∈ kH (or equally well k ∈ gH).
    gH = kH iff $g^{-1}k \in H$ (or equally well $k^{-1}g \in H$).

This also means that a coset is completely determined by any element in that coset. Sometime it is convenient to talk about a coset by focusing on a single arbitrary but fixed element in the coset. That fixed element of the coset that we choose to represent the whole coset is called the *coset representative*, plainly enough. It will frequently be a step in a proof to choose a complete set of coset representatives, one from each coset. Let's next consider two examples.

In Q_8 take $H = <a> = \{1, a, a^2, a^3\}$. Refresh your memory of how the group operation works in Q_8 if necessary. Then $bH = \{b, ba, ba^2, ba^3\} = \{b, c, b^3, c^3\}$ and $Hb = \{b, ab, a^2b, a^3b\} = \{b, c, b^3, c^3\}$, which turns out to be the same set as bH though we end up writing the elements of the coset in a different order. Usually, however, the left coset will not be the same as the corresponding right coset when the group is non-Abelian. We will return to this later and determine exactly when the left and right cosets do coincide.

A more interesting example is D_6. In D_6 let's take H as one of the Klein 4-groups; suppose we choose $H = \{\rho_0, \rho_3, \delta_1, \mu_3\}$. There is some opportunity for confusion since we are using ρ both to indicate a rotation in D_6 and to indicate the right multiplication function, but the context should always make clear which one we mean. As we begin to determine all the cosets, the first one we should write down is the one whose representative is ρ_0, the identity element. This coset is just H itself. We could have taken any element of H to be the representative but we will almost always take the identity element as the representative of the subgroup itself. To find another coset we need not consider the four elements of H any longer in our search; using any of the elements of H as the representative will just give us H again. To find another coset, use one of the eight remaining elements, and it does not matter which one. Arbitrarily, I choose δ_2 as the representative for the second coset, and compute $\delta_2 H$. I will compute the right cosets after I compute the left cosets.

$\delta_2 H = \{\delta_2\rho_0, \delta_2\rho_3, \delta_2\delta_1, \delta_2\mu_3\} = \{\delta_2, \mu_1, \rho_2, \rho_5\}$

We need not consider any of these four elements any longer in our search for another coset; again any one of them could have represented this particular coset. There are only four elements left and if they don't form a coset then we have made a mistake somewhere. As the final representative, let's choose δ_3:

$\delta_3 H = \{\delta_3\rho_0, \delta_3\rho_3, \delta_3\delta_1, \delta_3\mu_3\} = \{\delta_3, \mu_2, \rho_4, \rho_1\}$

These are the left cosets. We can compute the right cosets in a similar fashion, but we may be forced to choose different representatives because the cosets themselves will be different sets. The coset represented by the identity is always H, so we are free to choose δ_2 to represent the second right coset:

$H\delta_2 = \{\rho_0\delta_2, \rho_3\delta_2, \delta_1\delta_2, \mu_3\delta_2\} = \{\delta_2, \mu_1, \rho_4, \rho_1\} \neq \delta_2 H$

Since δ_3 is not in $H\delta_2$ we are free to choose it as the representative to determine our third coset:

$H\delta_3 = \{\rho_0\delta_3, \rho_3\delta_3, \delta_1\delta_3, \mu_3\delta_3\} = \{\delta_3, \mu_2, \rho_2, \rho_5\} \neq \delta_3 H$

Hence the partition of left cosets is distinct from the partition of the right cosets, and this is very common with non-Abelian groups. For Abelian groups, of course, the left coset always equals the corresponding right coset.

A word of warning is in order here. Some authors, notably the late and great Marshall Hall write cosets with the opposite notation, in which Hg is a left coset rather than a right coset as I define it here, and gH is the right coset rather than the left. I don't know how much variation there is in the terminology from author to author. Let the reader beware.

Exercise 35: Prove Theorem 51 parts b), c), and d).

Exercise 36: In S_4 find the left cosets and the right cosets of $H = \{(1), (1, 2, 4), (1, 4, 2)\}$

Exercise 37: In S_4 let $H = \{(1), (1, 2, 4, 3), (1, 4)(2, 3), (1, 3, 4, 2)\}$ and find the left cosets and the right cosets of H.

Exercise 38: In U_{21} find the cosets of $H = <16>$.

Exercise 39: Let $G \simeq H \times K$. Find all the cosets of H in G.

Exercise 40: If $Hg = g'H$ for some g' distinct from g show that in fact $Hg = gH$.

Lesson 12: Lagrange's Theorem

We now arrive at the payoff for the work we have done in these first eleven lessons. Once we have learned to count things, any things, then we have gained the key to knowledge of them. We have long been presumptuous enough to tacitly believe the next theorem, though we have never used it in any proof. Now we need be presumptuous no longer.

Theorem 52: (Lagrange's Theorem) Let G be a group of order n and let $H \leq G$ have order m. Then $m \mid n$.

Proof: The cosets of H form a partition of G and we begin by choosing a complete set of coset representatives: $\{1, g_1, g_2, \ldots, g_k\}$. Then $G = H \cup g_1H \cup g_2H \cup \ldots \cup g_kH$ because every element of G is in some coset. Since each coset has the same number of elements, and since no two cosets intersect each other, $n = |G| = \underbrace{m + m + \ldots + m}_{k+1 \text{ times}} = (k+1) \cdot m$. Therefore $m \mid n$.

Unfortunately, the converse of Lagrange's Theorem is not true. The order of a group can be divisible by a number and yet there may be no subgroup of that order. One example we already know of this is $\mathbf{A_4}$ which has order 12. 12 is divisible by 6 but $\mathbf{A_4}$ has no subgroup of order 6, as you should confirm. We will come to a partial converse to Lagrange's Theorem before long. One easy corollary of Lagrange's Theorem has long been believable but now can be firmly established:

Corollary 52a: Let p be a prime. Up to isomorphism, the only group of order p is $\mathbf{Z_p}$.
Proof: is an easy exercise.

Corollary 52b: Let G be a group of order n and let $g \in G$. Then the order of g divides the order of G.
Proof: is an easy exercise.

It will be useful to add to our terminology a bit here.

Definition 36: Let G be a group and $H \leq G$. The number of left (or right) cosets of H in G is called the **index of H in G** and is denoted by [G:H].

Corollary 52c: $|G| = [G:H] \cdot |H|$
Proof: another easy exercise.

We will look at four more theorems to finish this lesson and illustrate the value of Lagrange's Theorem.

Theorem 53: Let $H \leq K \leq G$. Then $[G:H] = [G:K] \cdot [K:H]$.

Proof: First we will list the left cosets of H in K. Choose the representatives of these cosets as $\mathbf{1}, k_1, k_2, \ldots, k_{n-1}$, noting that all these representatives are elements of K. Thus the index of H in K is n and there are exactly n cosets of H in K. Then $K = H \cup k_1H \cup k_2H \cup \ldots \cup k_{n-1}H$. Now do the same thing, enumerating the cosets of K in G. The representatives of these cosets will be taken as $1, g_1, g_2, \ldots, g_{m-1}$. We note that each of these representatives are elements of G that are not elements in K except for $\mathbf{1}$ of course. Thus the index of K in G is m, and we then have $G = K \cup g_1K \cup g_2K \cup \ldots \cup g_{m-1}K$. Substitute the union for K into the union for G:

$G = (H \cup k_1H \cup k_2H \cup \ldots \cup k_{n-1}H) \cup g_1(H \cup k_1H \cup k_2H \cup \ldots \cup k_{n-1}H)$
$\cup g_2(H \cup k_1H \cup k_2H \cup \ldots \cup k_{n-1}H) \cup \ldots \cup g_{m-1}(H \cup k_1H \cup k_2H \cup \ldots \cup k_{n-1}H)$.

This expresses G as the disjoint union of $n \cdot m$ cosets of H, so we know $[G:H] \leq [G:K] \cdot [K:H]$ but the question remains: are these cosets of H all distinct? Choose two distinct representatives of the cosets of K, say g_i and g'_i, and two distinct representatives of the cosets of H, say k_j and k'_j and suppose $g_ik_jH = g'_ik'_jH$. Then $g_i(k_jH) = g'_i(k'_jH)$. But the cosets k_jH and k'_jH are both contained in K. So by theorem 49a $g_iK = g'_iK$. Then g_i and g'_i do not represent distinct cosets of K, contradicting our choice. Hence we have written G as $n \cdot m$ disjoint cosets of H, and the theorem is proved.

Here is a somewhat technical result which we will need occasionally in future proofs.

Theorem 54: Let $H, K \leq G$. Then $[H \vee K:H] \geq [K:K \cap H]$. Further $[H \vee K:K] = [K:K \cap H]$ only if $[H \vee K:K]$ and $[H \vee K:H]$ are relatively prime.

Proof: To make our notation more convenient, let $J = K \cap H$. Choose representatives for the cosets of J in K, say $1, k_1, k_2, \ldots, k_{n-1}$. Then the cosets $J, k_1 J, k_2 J, \ldots, k_{n-1} J$ are all distinct in K. We claim that the cosets $H, k_1 H, k_2 H, \ldots, k_{n-1} H$ are also distinct in $H \vee K$. For suppose $k_i H = k_j H$. Then by theorem 49d, $k_i = k_j h$ for some $h \in H$. But then $k_j^{-1} \cdot k_i = h$. Hence $k_j^{-1} \cdot k_i$ is an element of both H and K. Hence $k_j^{-1} \cdot k_i \in J$ and $k_i \in k_j J$, contradicting the way we chose the k's. Therefore the cosets $H, k_1 H, k_2 H, \ldots, k_{n-1} H$ are all distinct in $H \vee K$ and therefore the index of H in $H \vee K$ is at least as large as the index of $K \cap H$ in K.

To prove the second statement, we assume $[H \vee K:K]$ and $[H \vee K:H]$ are relatively prime and use theorem 53 twice:
$$[H \vee K:K \cap H] = [H \vee K:K] \cdot [K:K \cap H] = [H \vee K:H] \cdot [H:K \cap H]$$
By the first statement of this theorem $[H \vee K:K] \geq [H:K \cap H]$. But since, by assumption, $[H \vee K:K]$ is relatively prime to $[H \vee K:H]$, $[H \vee K:K]$ must divide $[H:K \cap H]$. Thus $[H \vee K:K] = [H:K \cap H]$.

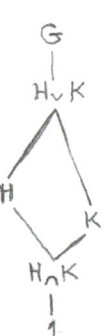

In the lattice diagram given here, which is meant to illustrate this theorem, I have made the lengths of the line segments connecting two subgroups to be proportionalish to the index between them. The lattice diagram is not for a specific group, this time. It is for a group in general. We are moving to the phase of study in which we seek to understand groups that are too large to grasp in detail. Hence the lattice diagram is limited to the subgroups under consideration; you must take for granted that there are many other subgroups but only the relevant ones are being shown.

There is a generalization of the coset which will occasionally be helpful.

Definition 37: Let A and B be any two subsets of the group G. The set
$$AB = \{ab \mid a \in A, b \in B\}$$
is called a **complex**. If A and B are both subgroups of G, then the set
$$AgB = \{agb \mid a \in A, b \in B\}$$
is called a **double coset**.

Typically, a complex will not itself be a subgroup. If we choose B to be a subgroup and we choose $A = \{a\}$, then the complex AB is just the left coset aB. Thus left or right cosets are special cases of a complex. More generally, a complex is just the set of all the products in a group that have a particular form specified by the name of the complex. The expression bCdE will mean the set of all the possible elements of the group which have the form bcde where b and d are fixed elements, c can be any element of the set C, and e can be any element of the set E. Most commonly, we will be interested in complexes of the forms AB and AgB.

Note that the double coset AgB can be thought of as the union of all the right cosets of A whose representatives have the form gb for some $b \in B$. Or, equally well, AgB can be thought of as the union of all the left cosets of B whose representatives have the form ag for some $a \in A$.

We are interested in knowing two things in particular about complexes. First, when is a complex a subgroup? And second, how big is a complex. Here are the answers:

Theorem 55: Let H and K be subgroups of G. Then $HK \leq G$ iff $HK = KH$.

Proof: Note that HK is the set of all elements of the form hk and KH is the set of all elements of the form kh so that in a non-Abelian group there is no reason to suppose that the two complexes are the same.

$\Rightarrow$) Suppose $HK \leq G$ and let hk be an arbitrary element of HK. Then $(hk)^{-1} = k^{-1}h^{-1} \in HK$. Since inverses are unique, and each element of HK is the inverse of some element of HK, each element of HK is also an element of KH, HK $\subseteq$ KH. Similarly if kh $\in$ KH, then kh is the inverse of $h^{-1}k^{-1} \in$ HK and hence kh $\in$ HK and KH $\subseteq$ HK. Therefore HK = KH.

$\Leftarrow$) Suppose HK = KH. We immediately know **1** $\in$ HK. Now if hk $\in$ HK, then we also know that $(hk)^{-1} = k^{-1}h^{-1} \in$ KH = HK. If h_1k_1 and $h_2k_2 \in$ HK, and consider $h_1k_1h_2k_2 = h_1(k_1h_2)k_2$. Since $k_1h_2 \in$ KH and KH = HK, there exist some elements h_3, k_3 such that $k_1h_2 = h_3k_3$. Hence we can compute $h_1k_1h_2k_2 = h_1h_3k_3k_2 = (h_1h_3)(k_3k_2) \in$ HK. Therefore HK $\leq$ G.

Be careful not to think that HK = KH means that the elements of H commute with the elements of K. The left hand side need not have the same elements of H and K as the right.

Theorem 56: Let G be a group; let H and K be subgroups of G. Then $|HK| = (|H|\cdot|K|) / |H \cap K|$
Proof: is left as an exercise.

Exercise 41: Confirm that $\mathbf{A}_4$ has no subgroup of order 6. (Hint: first show that there are only two possible groups, up to isomorphism, of order 6.)
Exercise 42: Prove Corollaries 52a, 52b and 52c.
Exercise 43: Prove Theorem 56.
Exercise 44: Prove that if n is odd, then $\mathbf{D}_n$ has no subgroup isomorphic to **V**.
Exercise 45: In $\mathbf{D}_6$ find the complex $\mathbf{V}\mathbf{Z}_3$ choosing any copy of **V**. Find the complex $\mathbf{V}_1\mathbf{V}_2$ choosing any two distinct copies of **V**.

> Joseph-Louis Lagrange was born in Turin, Italy, in 1736, to a French army officer and an Italian woman. He spent his whole childhood in Italy and was intended by his father to study law. At the university in Turin he dutifully studied law, though his favorite class was Latin. At 17 he encountered a paper by Edmund Halley and became enraptured by mathematics. His entire mathematical education was self-taught. His reputation was so great that in 1766 he was invited to Berlin to be Euler's successor. He had already been in long correspondence with Euler and had done research of the highest level with him. In 1783 his wife Vittoria died and he went into a long depression; much of his most famous work was done before 1783. In 1786 he moved to Paris, just as the political climate was deteriorating. He survived the Reign of Terror at Lavoisier's intervention, though Lavoisier himself was eventually guillotined. Later he was a favorite of Napoleon's. He died in 1813. With Euler he derived the Euler-Lagrange equations and founded calculus of variations. This he used with the principle of least action to reformulate Newtonian physics into what is now called Lagrangian mechanics. His analysis of orbits uncovered the Lagrange points. He developed the method of Lagrange multipliers and variation of parameters, which are still studied in differential equations. Though theorem 52 is named for Lagrange, he never proved it; he only stated a special case of it in 1771. Gauss extended it in 1801, then Cauchy extended it further in 1844. It was Camille Jordan who finally proved it in 1861, still before the concept of a group had been formally defined. Lagrange's greatest work was the book *Mechanique analytique*, which he was revising when he died.

Lesson 13: Normal Subgroups

We are now ready to introduce one of the most fruitful ideas in Modern Algebra. All the rest of this particular book will be the study of normal subgroups and their properties and implications.

To motivate the definition somewhat, consider the problem of studying non-Abelian groups. The Abelian groups were relatively easy, but the non-Abelian groups are a great deal more diverse and complicated. In order to get some handle on them, there are several methods that are used but at root they each amount to trying to make the non-Abelian groups as nearly Abelian as possible. If we can't use the commutative law, are there are situations when the commutative law *almost* applies? The answer will clearly depend on what we mean by "almost".

Our first foray into this approach will be based on the idea of the coset. We noted that for Abelian groups, whatever the subgroup we begin with, the left coset is always the same as the corresponding right coset: $gH = Hg \; \forall g \in G$. Does this reversal of left and right on the coset level ever occur in non-Abelian groups even though the operation itself is non-commutative? As it turns out, the answer is yes and this leads us to the definition.

Definition 38: Let G be a group and $H \leq G$. If $gH = Hg \; \forall g \in G$, then we say that H is a **normal subgroup** of G, and we write $H \triangleleft G$.

It is important to realize what is not true about normal subgroups. Except for the identity element, g may not actually commute with any element of H at all. It commutes with the *set of elements* only. In other words, we have the general rule:

If $H \triangleleft G$, then for every $g \in G$ and for every $h \in H$ there exists some $h' \in H$ such that $gh = h'g$.

We can always reverse the order of the product but we may have to switch to a different element of H to do it. The elements of the normal subgroup *almost* commute with the elements of the group.

But is there any easy way to tell if a subgroup is normal or not? Sometimes there is. The next two theorems reveal two situations in which we can be sure a subgroup is normal.

Theorem 57: Let G be a group and let $H \leq G$. If $[G:H] = 2$ then $H \triangleleft G$.

Proof: That the index of H in G is 2 means that H has exactly two cosets, including H itself, of course. In terms of left cosets, this means that $G = H \cup gH$ for g not an element of H; and in terms of right cosets it means $G = H \cup Hg$. Thus both gH and Hg are the set complement of H, since they are both disjoint from H. Therefore $gH = Hg$.

This theorem has two immediate corollaries concerning the groups we studied in part 1, and leads immediately to another theorem.

Corollary 57a: For all $n > 2$, $A_n \triangleleft S_n$.

Proof: Compare the orders of the two groups.

Corollary 57b: For all $n > 2$, $Z_n \triangleleft D_n$.

Proof: Compare the order of the dihedral group to the order of its subgroups of rotations.

Theorem 58: Let $G = H \times K$ be a group. Then $H \triangleleft G$ and $K \triangleleft G$.

Proof: this is left as an exercise.

Unfortunately, normal subgroups do not behave in as convenient a fashion as characteristic subgroups do. We proved that the characteristic subgroup of a characteristic subgroup is a characteristic subgroup; the relation of being a characteristic subgroup is a transitive relation. This is not the case for normal subgroups. *The normal subgroup of a normal subgroup need not be a normal subgroup*. In the exercises you will be asked to show that A_4 provides us with a counter-example. Life would be simpler – but less interesting - if normality were transitive. Nearly all the rest of this book hinges directly or indirectly on the fact that normality fails to be transitive. There is, however, one situation in which we can count on normality being transitive.

Theorem 59: Let $G = H \times K$ be a group. If $S \triangleleft H$, then $S \triangleleft G$.

Proof: Assume $G = H \times K$ and that $S \triangleleft H$. Choose any element of G which is not in H, say (h, k) where $k \neq 1$. Any element of S has the form $(s_i, 1)$ for some $s_i \in S$. Thus any element in the left coset $(h, k)S$ has the form $(h, k)(s_i, 1) = (h \cdot s_i, k)$. But since $S \triangleleft H$ there is some element of S, say s_j, such that $h \cdot s_i = s_j \cdot h$. Thus $(h \cdot s_i, k) = (s_j \cdot h, k) = (s_j, 1)(h, k) \in S(h, k)$. Since we began with an arbitrary element of S, we have shown $(h, k)S = S(h, k)$. Therefore, since we also began with an arbitrary element of G, $S \triangleleft G$.

Though normal subgroups give a kind of almost commutativity, when we have two normal subgroups under certain circumstances, we get full commutativity.

Theorem 60: Let G be a group, $H \triangleleft G$, $K \triangleleft G$ and such that $H \cap K = 1$. Then for any $h \in H$ and any $k \in K$, $hk = kh$.

Proof: We suppose $H \triangleleft G$, $K \triangleleft G$ and $H \cap K = 1$, and consider the product hk. First, because $K \triangleleft G$, we know there is some element $k' \in K$ such that $hk = k'h$. Also because $H \triangleleft G$, we know there is some element $h' \in H$ such that $hk = kh'$. Therefore $kh' = k'h$. This means that we have $h'h^{-1} = k^{-1}k'$. Thus $h'h^{-1}$ and $k^{-1}k' \in H \cap K = \mathbf{1}$. Hence $h'h^{-1} = 1$ and $k^{-1}k' = 1$. This gives $h' = h$ and $k = k'$. Therefore $hk = kh$ and the theorem is proved.

It is important to notice here that while elements of H will commute with elements of K, elements of H may not commute with each other.

Theorem 61: Let G be a group; let H and K be subgroups of G. If either H or K (or both) is normal in G, then $H \vee K = HK$.

Proof: Recall that $H \vee K$ is the smallest subgroup containing both H and K. Clearly $HK \subseteq H \vee K$. Hence we need only prove that HK is a subgroup of G. By theorem 55, HK is a subgroup of G iff $HK = KH$. Let $h_1k_1 \in HK$ be arbitrary, and without loss of generality, let's suppose it is H that is normal in G. Then there is some $h_2 \in H$ such that $h_1k_1 = k_1h_2 \in KH$. Hence $HK \subseteq KH$. The same argument beginning with $k_1h_1 \in KH$ shows that $KH \subseteq HK$. Therefore $HK = KH$. By theorem 55, HK is a subgroup and must equal $H \vee K$.

We will end this lesson with one final connection between normal subgroups and direct products. This theorem extends theorem 41, which gave one way of discerning when a finite Abelian group was the internal direct product of two of its subgroups and showed one way of telling which subgroups were the factors. We can now extend this result to non-Abelian groups. Non-Abelian groups are seldom internal direct products but theorem help to discern when they are.

Theorem 62: Let G be a group and suppose G has two normal subgroups, H and K, with $H \vee K = G$ and $H \cap K = 1$. Then $G \simeq H \times K$.

Proof: By the previous theorem, $H \vee K = HK = G$. We proceed as usual by proposing a candidate for the isomorphism. Let $\phi: H \times K \to G$ be defined by the following: $\phi((h, k)) = hk$. Is ϕ an injection? Suppose $\phi((h, k)) = \phi((h', k'))$. Then by definition, $hk = h'k'$. This implies that $kk'^{-1} = h^{-1} \cdot h'$. But this means that $h^{-1} \cdot h' \in H \cap K = \mathbf{1}$. Hence $h^{-1} \cdot h' = 1$ and thus $h' = h$. Then $hk = hk'$ and so $k = k'$. Thus we have shown that ϕ is a injection.

Since these are finite groups, if ϕ is an injection, then ϕ must also be a surjection, but it is also clear because $G = H \vee K = HK$ and so every element of G has the form hk and is thus the image of (h, k) under ϕ. Thus ϕ is a bijection.

Does ϕ preserve products? Compute $\phi((h, k) \cdot (h', k'))$ in both ways:
$\phi((h, k) \cdot (h', k')) = \phi((hh', kk')) = hh'kk'$.

$\phi(\,(h, k)\,) \cdot \phi(\,(h', k')\,) = hk \cdot h'k' = hh'kk'$ by theorem 60.

Therefore ϕ preserves products and is an isomorphism.

Compare this theorem with theorem 50. We will see more clearly how theorems 50 and 62 are related shortly. There is a great deal more to be said about the properties of normal subgroups, but we will wait to say it until we learn how to discuss normal subgroups from an entirely different viewpoint. Here we have analyzed normal subgroups using their cosets. Next we will consider them from the vantage point of a special kind of automorphism.

Exercise 46: Show explicitly, using A_4, that if $H \triangleleft K$ and $K \triangleleft G$, then H need not be normal in G.

Exercise 47: In D_6 there are two distinct subgroups isomorphic to D_3. Show that they are both normal in D_6. Show that the elements in one copy of D_3 do not necessarily commute with the elements of the other copy of D_3.

Exercise 48: Let $H \triangleleft G$ and $J \triangleleft K$ and show that $H \times J \triangleleft G \times K$.

Exercise 49: Prove theorem 58. Is it true that if $G = H \times K$ then $H \ll G$ and $K \ll G$?

Exercise 50: Let $H \triangleleft G$ and $[G:H] = p$, a prime. Let K be any subgroup of G. Show that either $K \leq H$; or $G \simeq HK$ and $[K : K \cap H] = p$.

Lesson 14: Conjugations and Inner Automorphisms

There is a second, equivalent way of defining normal subgroups, and it is important to be able to think of them from either viewpoint. This second viewpoint approaches normal subgroups by way of special automorphisms. They are closely related to the left and right multiplication maps, and are another way of investigating exactly how far we can push commutativity in non-Abelian groups.

Definition 39: Let G be a group and $g \in G$ any element of the group. The map $\kappa_g: G \to G$ defined by $\kappa_g(x) = g^{-1}xg$ is called **conjugation by g**. If $g^{-1}xg = y$, we say that x and y are **conjugates** of each other.

It is clear that if G is an Abelian group, every conjugation map simply collapses to the identity map. Thus conjugation maps are designed for the study of non-Abelian groups. When x and y are conjugate elements under κ_g then they "almost commute" with the element g. The conjugation map looks at commutativity on the elemental level rather than on the coset level. The first theorem tells us that κ_g is much easier to deal with than λ_g or ρ_g.

Theorem 63: Let G be a group and $g \in G$. Then κ_g is an automorphism of G.

Proof: It is immediate from the definition of κ_g that it is a bijection from G onto G. We need only show that κ_g preserves products. So compute the conjugate of a product and compare it to the product of the conjugates:

$$\kappa_g(xy) = g^{-1} \cdot xy \cdot g \text{ by definition of } \kappa_g$$
$$\kappa_g(x) \cdot \kappa_g(y) = (g^{-1} \cdot x \cdot g) \cdot (g^{-1} \cdot y \cdot g) = g^{-1} \cdot x \cdot (g \cdot g^{-1}) \cdot y \cdot g = g^{-1} \cdot xy \cdot g$$

Hence $\kappa_g \in \mathbf{Aut}(G)$.

Thus, by theorem 37e, if $H \leq G$, $\kappa_g(H) \leq G$. So if $\kappa_g(H) = K$, we will naturally say that H and K are *conjugate subgroups* of each other. It is easy to show the following:

Theorem 64: $H \triangleleft G$ iff $\kappa_g(H) = H$ for every $g \in G$.

Proof: ⇒) Suppose $H \triangleleft G$. Then $Hg = gH$ for any $g \in G$, and therefore $g^{-1}Hg = g^{-1}gH = H$, which is the same as $\kappa_g(H) = H$ for any $g \in G$.

⇐) Now suppose $\kappa_g(H) = H$. This is $g^{-1}Hg = H$ which implies $Hg = gH$. Hence $H \triangleleft G$.

Thus, normal subgroups are those subgroups that are preserved by every conjugation map, and every conjugation map is an automorphism; but not conversely. There are automorphisms that are not conjugations by any element. For example, *every* automorphism of an Abelian group is not a conjugation map since for Abelian groups conjugation maps are always the identity map. This does lead us to new terminology.

Definition 40: Let G be a group and let $\beta \in \mathbf{Aut}(G)$. If there exists some element $g \in G$ such that $\beta = \kappa_g$ then β is called an **inner automorphism** of G. We will use the notation **Inn**(G) to denote the set of all the inner automorphisms of G.

We will define outer automorphisms soon so that the meaning of "inner" will not be left dangling. This leads to an alternative way of defining normal subgroups.

Definition 38': Let $H \leq G$ and suppose $\kappa_g(H) = H$ for all inner automorphisms $\kappa_g \in \mathbf{Aut}(G)$. Then we say that H is a **normal subgroup** of G.

The two definitions are equivalent; either will do perfectly well. Now let's explain the reason behind the terminology. The idea is that conjugation maps are automorphisms that arise from within G itself, whereas an automorphism that is not a conjugation by some element comes from somewhere outside of G. We will return to this idea when we have more tools to examine it. Thus there is a relationship between characteristic subgroups and normal subgroups. Characteristic subgroups are preserved by every automorphism; normal subgroups are preserved only by the inner automorphisms. The next theorem is immediate.

Theorem 65: Let G be a group. If H ≪ G then H ◁ G.
Proof: is immediate.

But not conversely. A normal subgroup is not necessarily preserved by any non-inner automorphism. In an Abelian group, every inner automorphism is just the identity map and so it is trivial that every subgroup is preserved by every inner automorphism and so every subgroup of an Abelian group is normal. But we noticed that in some Abelian groups some of these normal subgroups were not preserved by all automorphisms and hence were not characteristic.

There is more to the relationship between characteristic subgroups and normal subgroups. For one thing, we can now see more clearly how theorems 50 and 62 are related. Characteristic subgroups of characteristic subgroups are characteristic; normal subgroups of normal subgroups may not be normal. What about characteristic subgroups of normal subgroups? Or normal subgroups of characteristic subgroups?

Theorem 66: Let G be a group, let K ◁ G, and let H ≪ K. Then H ◁ G.

Proof: Let κ_g be any inner automorphism of G. Since K ◁ G, $\kappa_g(K) = K$. Therefore $\kappa_g|_K$ is an automorphism of K. $\kappa_g|_K$ is not an inner automorphism of K because the conjugating element g need not be an element of K. However, since H ≪ K, $\kappa_g|_K(H) = H$ regardless of whether $\kappa_g|_K$ is inner or not. Therefore $\kappa_g(H) = H$ for any $g \in G$ and so H ◁ G.

Again, the converse of theorem 66 is not true. *A normal subgroup of a characteristic subgroup may not be normal.* In the exercises you will be asked to find an example of this fact. Now we will prove that, not only do the inner automorphisms of G form a subgroup of the group of all the automorphisms, but in fact they form a *normal subgroup* of the group of all the automorphisms.

Theorem 67: Let G be a group. Then **Inn**(G) ◁ **Aut**(G).

Proof: First it is clear that the identity automorphism is a conjugation map, an inner automorphism.

Now suppose κ_g is any inner automorphism. Let $\kappa_g(x) = g^{-1} \cdot x \cdot g = y$. Then it is immediate that $x = g \cdot y \cdot g^{-1} = (g^{-1})^{-1} \cdot y \cdot g^{-1} = \kappa_{g^{(-1)}}(x)$. In other words the inverse map to conjugation by g is conjugation by g^{-1}, and so the inverse of an inner automorphism is an inner automorphism.

Are products of inner automorphisms still inner automorphisms? It is easy to check:
$$(\kappa_g \circ \kappa_h)(x) = \kappa_g(\kappa_h(x)) = \kappa_g(h^{-1}xh) = \kappa_g(h^{-1})\kappa_g(x)\kappa_g(h) = (g^{-1} \cdot h^{-1} \cdot g)(g^{-1} \cdot x \cdot g)(g^{-1} \cdot h \cdot g)$$
$$(g^{-1} \cdot h^{-1})(gg^{-1})x(gg^{-1})(hg) = (g^{-1} \cdot h^{-1})x(hg) = (hg)^{-1}x(hg) = \kappa_{hg}(x)$$

Hence the composition of two conjugations is the conjugation by their product. This shows that **Inn**(G) ≤ **Aut**(G).

To show that **Inn**(G) ◁ **Aut**(G) requires more care than usual. We must show that **Inn**(G) is preserved by every inner automorphism of **Aut**(G); or, with other words, we must show that conjugation of an inner automorphism of G by an arbitrary automorphism of G yields another inner automorphism of G. So let κ_g be an arbitrary inner automorphism of G, and let β be an arbitrary outer automorphism of G. We need to show that $\beta^{-1} \circ \kappa_g \circ \beta$ is an inner automorphism of G. Let x be an arbitrary element of G. Then:
$$(\beta^{-1} \circ \kappa_g \circ \beta)(x) = (\beta^{-1}(\kappa_g(\beta(x))) = \beta^{-1}(g^{-1} \cdot \beta(x) \cdot g) = \beta^{-1}(g^{-1}) \cdot \beta^{-1}(\beta(x)) \cdot \beta^{-1}(g)$$
$$= (\beta^{-1}(g))^{-1} \cdot x \cdot \beta^{-1}(g)$$

Hence if we conjugate κ_g by β we get conjugation by $\beta^{-1}(g)$. Therefore **Inn**(G) ◁ **Aut**(G).

There is another notation that is increasingly used for conjugation. We will sometimes write the conjugation by g in *exponential notation* with g as the exponent. So $\kappa_g(h) = h^g$.

Exercise 51: Is it true that **Inn**(G) ≪ **Aut**(G)? Care is needed on this one. This will involve working in the group **Aut**(**Aut**(G)).

Exercise 52: Find **Inn**(D_4).

Exercise 53: Find **Inn**(D_5).

Exercise 54: Find an example that shows that a normal subgroup of a characteristic subgroup need not be normal.

Exercise 55: Express the composition of two conjugation maps in exponential notation.

Lesson 15: Further Properties of Normal Subgroups

The theorems we consider in this lesson are what might be called "technical details". They are not so interesting in themselves, but they are quite useful things to know in certain instances. Normality is not transitive in general, but it does *transfer* from one pair of subgroups to another pair of subgroups on occasion; and it is always useful to know when it does. Normality can be essential to know even on a relative scale. In other words, sometimes we do not need a subgroup to be normal in the whole group; sometimes it is enough to know that a subgroup is normal in some other subgroup containing it. The following fact is frequently helpful.

Theorem 68: If G is a group, H ◁ G, and K ◁ G. Then **(a)** H ∩ K ◁ G and **(b)** H ∨ K ◁ G.

Proof: We will prove the second statement here and leave the first statement as an exercise. Since H and K are normal subgroups, by theorem 61 we know H ∨ K = HK. Now let κ_g be any inner automorphism of G, and consider κ_g(H ∨ K) = κ_g(HK). If we let hk be any element of HK, then κ_g(hk) = κ_g(h)κ_g(k). Since H and K are normal, we know κ_g(h) = h' ∈ H and κ_g(k) = k' ∈ K, so κ_g(h)κ_g(k) = h'k' ∈ HK. Hence HK is preserved by every inner automorphism and so by definition HK = H ∨ K ◁ G.

The proof of part a) is left as an exercise.

The next theorem also allows us to find another normality relationship in a more general context. Note that if H is normal in the subgroup K containing H but not in the whole group G, this means that every inner automorphism by elements of K will preserve H, but that there are inner automorphism by elements of G not in K that do not preserve H. It will become critical in part 3 of Modern Algebra to be able to tell when one subgroup is normal in another subgroup. It is a consequence of the non-transitivity of normality that this complication arises at all. However, it is clear that if H is normal in the whole group G, then it is normal in any subgroup of G that contains it.

The next theorem is complicated enough that it merits a lattice diagram to picture it. The normality relations between the subgroups in the hypothesis of the theorem are shown in red, with a dotted line when the line is not necessary for the lattice diagram; the conclusion of the theorem is shown in green.

Theorem 69: Let G be a group and let H, K and S be subgroups of G such that both H ◁ S, and K ◁ G. Then H ∨ K ◁ S ∨ K.

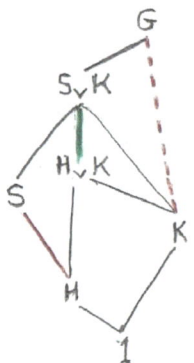

Proof: Since K ◁ G, we can conclude that H ∨ K = HK and S ∨ K = SK by theorem 61. We will need to show that HK is preserved by conjugation by any element of SK. All we need do is check:

(sk)⁻¹(hk)(sk) = k⁻¹s⁻¹hksk = k⁻¹s⁻¹h(ss⁻¹)ksk = k⁻¹(s⁻¹hs)(s⁻¹ks)k
= k⁻¹(h')(s⁻¹ks)k   (because H ◁ S)
= k⁻¹h'(k')k   (because K ◁ G and s ∈ G)
= k⁻¹h'k''k = h'k'''k''   (because K ◁ G)

and h'k'''k'' ∈ HK. Therefore HK ◁ SK.

Theorem 70: Let G be a group, K ◁ G, and H ≤ G. Then K ∩ H ◁ H.

Proof: is left to the exercises.

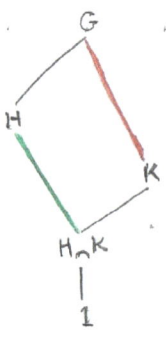

To repeat myself, it is always good to be able to count things. Knowing how many cosets a subgroup has was the key to Lagrange's theorem, which is invaluable in understanding groups. The next theorem is not quite so invaluable, but it does give us a way of counting the cosets of one group that lie inside a particular double coset. Before we look at the theorem, let's consider the nature of conjugate subgroups. A conjugate subgroup to the subgroup H is the image of H under an inner automorphism. Hence, a conjugate subgroup to H is isomorphic to H. Normal subgroups have no conjugate subgroups other than themselves, but any subgroup that is not normal must have at least one conjugate subgroup, and perhaps very many of them. With that said, we can look at the next theorem. The lattice diagram at the end may be helpful in following the details of the proof.

Theorem 71: Let G be a group and let H and K be any two subgroups of G. Then **(a)** the number of right cosets of H contained in HxK equals $[K: K \cap x^{-1}Hx]$ and **(b)** the number of left cosets of K contained in HxK equals $[x^{-1}Hx: K \cap x^{-1}Hx]$.

Proof: To prove part (a) we will first set up a bijection between elements of the double coset HxK and elements of the complex $x^{-1}HxK$. The most obvious way is: $hxk \leftrightarrow x^{-1}hxk$. This will also establish a bijection between the right cosets Hxk of H in HxK and the right cosets $(x^{-1}Hx) \cdot k$ in $x^{-1}HxK$. To save on notation, use $L = x^{-1}Hx$ and let $L \cap K = M$. So we have a bijection between the right cosets of H in HxK and the right cosets of L in $x^{-1}HxK$. Now let $r = [L:M]$ so we can write
$$L = M \cup Mx_2 \cup Mx_3 \cup \ldots \cup Mx_r,$$
where the $x_i \in L$ are a complete set of coset representatives of M in L. Then I claim K, Kx_2, Kx_3, …, Kx_r are the distinct right cosets of K in KL. They are distinct because if $Kx_i = Kx_j$ then $x_ix_j^{-1} \in K$; and then, since x_j^{-1} and $x_i \in L$, we have x_j^{-1}, x_i and $x_ix_j^{-1} \in M$, contrary to the way the x_i were chosen. They are complete because if there were another right coset of K in KL, its representative would produce another right coset of M in L contrary to our assumption. Thus every right coset of K in KL is of the form Ka where a has the form $a = mx_i$ for some $m \in M$. But $M \leq K$, so $Kmx_i = Kx_i$. Thus the number of right cosets of K in KL is $[L:M] = [x^{-1}Hx: K \cap x^{-1}Hx]$. Hence, by our original bijection, this equals the number of right cosets of H in HxK.

The proof of part (b) is similar and is left to the exercises.

Exercise 56: Prove theorem 68 part a).
Exercise 57: Prove theorem 70.
Exercise 58: Prove theorem 71 part b).
Exercise 59: Let G be a group and let H, K and S be subgroups of G such that $H \triangleleft S$ and $K \triangleleft G$. Is it true that $H \cap K \triangleleft S \cap K$?

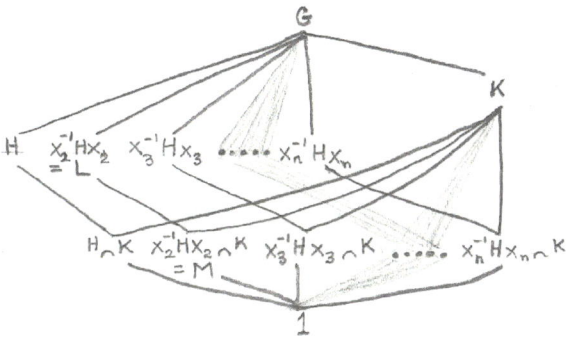

Lesson 16: Conjugation in S_n

It will be useful to have some facility in dealing with permutations. The symmetric groups S_n have particular interest, not only because they include all the permutations of n letters but because, as we shall see, they contain isomorphic copies of all groups. But they are large groups and we will need all the tools we can devise in order to understand them. In this lesson we will look at how easy it is to compute conjugations of permutations in these groups.

We will begin with a definition of a combinatorial concept:

Definition 41: Let N be any positive integer. A **partition of N** is any sequence of positive integers n_1, n_2, …, n_i such that $n_j \leq n_{j+1}$ for all $j < i$ and such that $\sum_j n_j = N$.

Thus a partition of N is any non-decreasing sequence of integers that add up to N. As examples, here are some partitions of 16:

1,1,1,1,1,1,1,1,1,1,1,1,1,1,1,1
1,1,1,2,2,2,2,2,3
1,2,3,4,6
16

If we ask, given the integer N, how many possible partitions of N are there, the question is not easily answered. That is a topic in the study of combinatorics and it would be a diversion of major proportions to pursue it here. We will use only the term. Closely related to the idea is the following:

Definition 42: Let $\sigma \in S_n$ be any permutation of n letters written as the product of disjoint cycles. Then the lengths of all the cycles written in non-decreasing order is called the **cycle type** of σ.

Clearly the cycle type of any permutation in S_n is a partition of n. It is also clear that the cycle type of any permutation is uniquely determined once we specify the number of letters involved. The cycle type of (1, 2, 4)(3, 5) is 2,3 if the permutation is understood as an element of S_5, but it has cycle type 1,2,3 if it is understood as an element in S_6. The ambiguity arises only because of our practice of not writing the fixed letters in cycle notation.

We now prove the theorem that is the primary object of this lesson.

Theorem 72: Two permutations are conjugate to each other iff they have the same cycle type.

Proof: $\Rightarrow$) Let σ and τ be any two permutations in S_n. Suppose σ, when written as the product of disjoint cycles, contains $(a_1, a_2,…, a_i)(b_1, b_2,…, b_j)$ with possible other cycles of different lengths. Consider the product $\tau\sigma\tau^{-1}$, which is conjugation by τ^{-1}. We conjugate by the inverse rather than by τ for reasons that will become clear. Let's see how the conjugation moves a letter $\tau(a_k)$. First τ^{-1} moves $\tau(a_k)$ to a_k; then σ moves a_k to a_{k+1}; finally τ moves a_{k+1} to $\tau(a_{k+1})$. Now we chose a_k arbitrarily. In other words, if σ moves some letter c to some letter d, then $\tau\sigma\tau^{-1}$ moves $\tau(c)$ to $\tau(d)$. Thus, $\tau\sigma\tau^{-1}$ will contain the cycles ($\tau(a_1), \tau(a_2),…, \tau(a_i)$)($\tau(b_1), \tau(b_2),…, \tau(b_j)$). The resulting cycles have the same lengths, but with the letters $\tau(a_i)$ in place of the letters a_i. So conjugate permutations have the same cycle type.

$\Leftarrow$) Now suppose σ_1 and σ_2 in S_n have the same cycle type. Choose a factorization into disjoint cycles for each of them including the 1-cycles; arrange these cycles according to length, from the shortest to the longest. This may be done in many ways since disjoint cycles commute with each other. In each of these products, the letters from 1 to n will appear exactly once. Now let τ be the function which maps the ith letter in the factorization for σ_2 to the ith letter in the factorization for σ_1. Then τ is a permutation of the n letters. Since τ is designed to preserve placement of the parentheses and merely substitute the letters of σ_2 for the corresponding letters of σ_1, τ^{-1} will substitute the letters of σ_2 for the corresponding letters of σ_1. The first half of this proof then shows that conjugation by τ^{-1} will change σ_1 into σ_2. They are therefore conjugate permutations.

As an example let's choose two permutations in S_9 with the same cycle type and find the conjugation map that will change the first to the second. Let

$\sigma_1 = (8)(2)(1, 5)(3, 4, 6, 7, 9)$
$\sigma_2 = (2)(7)(3, 8)(1, 9, 4, 6, 5)$

We want to find the permutation we have called τ^{-1} that changes the letters of σ_2 into the corresponding letters of σ_1 and conjugate by it. A little care following the letters tells us $\tau^{-1} = (2, 8, 5, 9, 4, 6, 7)(1, 3)$. Then $\tau = (1, 3)(2, 7, 6, 4, 9, 5, 8)$. So we compute the conjugation by τ^{-1}, $\tau\sigma_1\tau^{-1}$ and expect to get σ_2.

$(1, 3)(2, 7, 6, 4, 9, 5, 8)(8)(2)(1, 5)(3, 4, 6, 7, 9)(1, 3)(2, 8, 5, 9, 4, 6, 7)$
$= (1, 9, 4, 6, 5)(2)(3, 8)(7) = \sigma_2$

We should notice that the conjugating permutation, τ^{-1}, need not have the same cycle type as the permutations that are conjugate to each other. In this example, the cycle type of the σ's is 1,1,2,5 while the cycle type of τ is 2,7. In this case, all the permutations are odd permutations, but it may well happen that the σ's may be odd while the τ is even, or vice versa. Elements of A_n, which are all even, may be conjugate to each other, but only by an element that is not in A_n. Thus, in A_5 the 5-cycles (1, 2, 3, 4, 5) and (1, 3, 5, 2, 4) are not conjugate to each other; but there is an odd permutation, in S_5 but not in A_5, which makes them conjugate.

There is another result we should consider for the sake of completeness. We know that in S_n for any letter k we have a subgroup Stab(k) of all the permutations which fix k. Clearly Stab(k) is isomorphic to S_{n-1} and because there n letters we must have n such copies of S_{n-1}. What are the conjugates of a stabilizer? Let $\sigma \in$ Stab(k) and suppose we have a permutation that moves k to j; call it τ^{-1}. Then τ moves j to k. Now if we conjugate by τ to get $\tau^{-1}\sigma\tau$, we see that first τ moves j to k, then σ fixes k, and finally τ^{-1} moves k back to j. In short, the conjugate $\tau^{-1}\sigma\tau \in$ Stab(j). Hence the conjugate of a stabilizer is another stabilizer. We have shown the following:

Theorem 73: Let τ be a permutation that moves the letter j to the letter k. Then τ^{-1}*Stab*(k) τ = *Stab*(j).

Exercise 60: List all the partitions for each of the numbers 3, 4, and 6. Then find permutations that correspond to each of the cycle types you've found.

Exercise 61: Let $\sigma = (1, 4, 5, 2, 3, 6) \in S_6$. Find τ such that $\tau^{-1}\sigma\tau$ (conjugating by τ rather than by τ^{-1} as in the theorem) equals the following:

σ^2 $\qquad$ σ^3 $\qquad$ σ^{-1} $\qquad$ σ^{-3} $\qquad$ $\rho_1 = (1, 2, 3, 4, 5, 6)$ $\qquad$ $\rho_2 = (1, 6, 2, 5, 3, 4)$

Exercise 62: Write all the possible cycle types with 4 letters. Find an element of S_4 for each cycle type.

Lesson 17: Group Extensions

It is not unusual, when we are studying a particular group, that it is convenient to think of the group as a subgroup of a larger group with certain special properties. Discerning when such as expansion into a larger context is possible is not an easy problem in general, but in some cases it can be done. The distinction we have made between automorphisms which are inner and those which are "outer" (to be defined soon) leads naturally to that kind of question. Suppose β is an automorphism of the group G that is not inner. Is it possible to find another group $\check{G}$ such that G is isomorphic to a subgroup of $\check{G}$ and $\check{G}$ possesses an element x such that $\beta = \kappa_x|_G$. In other words, can we find a larger group such that the "outer" automorphism of our original group extends to an inner automorphism of the larger group? "Outer" automorphisms of G arise from outside of G. Can we find explicitly the outside group they arise from? There is a standard term that we can introduce here.

Definition 43: Let G be a group and $H \leq G$. We will call G an **extension** of H.

We are looking to find an extension $\check{G}$ of G in which a particular outer automorphism of G is a restriction of an inner automorphism of $\check{G}$. There is not a unique solution to this problem, but here we will try to find only the simplest possible extension of G with this property. Begin with the generators for G, say g is one of them. Let β be an outer automorphism of G. Then we know that $\beta(g)$ has the same order as g. Say $\beta(g) = h$. Assume $\check{G}$ exists and there is an $x \in \check{G}$ such that $\beta(y) = \kappa_x(y)$ for all $y \in G$. Then in particular $\beta(g) = h = \kappa_x(g) = x^{-1}gx$. Now β is a permutation of the elements of G and as a permutation β has a particular order. Let β have order m so that β^m is the identity automorphism of G. Then $\beta^m(g) = g = x^{-m}gx^m$. Because we are looking for the simplest extension, we will begin by assuming $x^m = 1$. In order to determine $\check{G}$ completely we need only find the relations that connect the new element x with all the generators of G.

The process is most clear by looking at a specific example. If we choose an Abelian group we know that all the automorphisms are outer ones, so lets begin with $\mathbf{V} = \{1, a, b, ab\}$ where we will, as usual, rename ab as c. There are two generators a and b. From lesson 7 let's consider the outer automorphism ϕ_3. In cycle notation, $\phi_3 = (a, b)$ and it has order 2. We will assume the existence of some element in some larger group, x, and we will assume x has order 2. Conjugation by x inside $\mathbf{V}$ matches ϕ_3 so we have the following:

$x^{-1}ax = b$ which means that $ax = xb$
$x^{-1}bx = a$ which means that $bx = xa$
$x^{-1}cx = c$ which means that $cx = xc$

Hence, if we throw x into the mix we generate a total of four new elements: x, ax, bx, and cx. We know the group extension we seek is non-Abelian (otherwise it would have no non-trivial inner automorphisms at all) but we know that the products in reverse order are determine when these four are determined. Consider the orders of ax, bx, and cx since we have taken x as being order 2:

$(ax)^2 = axax = abxx = c$ $(cx)^2 = cxcx = xccx = 1$
$(ax)^3 = cax = bx$
$(ax)^4 = c^2 = 1$

Thus cx is another element of order 2 but ax and bx are both elements of order 4. It is not a great imaginative leap to realize that our group extension is in fact $\mathbf{D_4}$. The copy of $\mathbf{V}$ we began with could be identified with either copy of the Klein 4-group in $\mathbf{D_4}$, so I will choose to identify a with μ_1 and b with μ_2. That identifies c with ρ_2 which is necessary because I can't use ρ_2 as a generator for either copy of $\mathbf{V}$ in $\mathbf{D_4}$. That ax is order 4 means that together a and x generate a copy of $\mathbf{D_4}$. The element ax could be identified with either ρ_1 or ρ_3 and that will be determined by whether we identify x with δ_1 or δ_2. Hence it is conjugation by either of the δ's in $\mathbf{D_4}$ that restricts to the automorphism ϕ_3 on that particular $\mathbf{V}$.

For another example, let's consider the outer automorphism of Q_8 which was given in lesson 9 as $\phi_{5,1} = (a, c, b)(a^3, c^3, b^3)$. It's order is 3 and so we will assume there is some extension with an element x of order 3 and we will force conjugation by x to equal $\phi_{5,1}$. We know that $a^4 = b^4 = c^4 = 1$ and that $a^2 = b^2 = c^2$ This will give us the following conditions on x to make it agree with $\phi_{5,1}$.

$x^{-1}ax = c$ which gives $ax = xc$
$x^{-1}cx = b$ which gives $cx = xb$
$x^{-1}bx = a$ which gives $bx = xa$
$x^{-1}a^3x = c^3$ which gives $a^3x = xc^3$
$x^{-1}c^3x = b^3$ which gives $c^3x = xb^3$
$x^{-1}b^3x = a^3$ which gives $b^3x = xa^3$
$x^{-1}a^2x = x^{-1}b^2x = x^{-1}c^2x = a^2 = b^2 = c^2$ (which means x commutes with the square of a, b or c.)

We need to find the orders of the elements x, ax, bx, cx, a^2x, a^3x, b^3x, c^3x, x^2, ax^2, bx^2, cx^2, a^2x^2, a^3x^2, b^3x^2, and c^3x^2, though we know already x and x^2 have order 3.

$(ax)^2 = axax = abxx = cx^2$ $(ax)^3 = axcx^2 = xccx^2 = a^2$ $(ax)^4 = a^2 ax = a^3x$
$(ax)^5 = a^3xax = a^3bxx = c^3x^2$ $(ax)^6 = c^3x^2ax = c^3xbxx = xb^3 bx^2 = 1$

Thus ax and c^3x^2 are order 6, cx^2 and a^3x are order 3 and this Z_6 shares the Z_2 in the original group. This leaves us to consider the elements bx, cx, a^2x, b^3x, c^3x, ax^2, bx^2, a^2x^2, a^3x^2, and b^3x^2.

$(bx)^2 = bxbx = bcxx = ax^2$ $(bx)^3 = bxax^2 = xaax^2 = a^2$ $(bx)^4 = a^2bx = b^3x$
$(bx)^5 = b^3xbx = b^3cxx = a^3x^2$ $(bx)^6 = a^3x^2bx = a^3xcxx = a^3 axx^2 = 1$

So bx and a^3x^2 are order 6, ax^2 and b^3x are order 3 and this Z_6 shares the Z_2 in the original group. We are now left to consider the elements cx, a^2x, c^3x, bx^2, a^2x^2, and b^3x^2.

$(cx)^2 = cxcx = caxx = bx^2$ $(cx)^3 = cxbx^2 = xbbx^2 = a^2$ $(cx)^4 = a^2cx = c^3x$
$(cx)^5 = c^3xcx = c^3axx = b^3x^2$ $(cx)^6 = b^3x^2cx = b^3xaxx = b^3bxx^2 = 1$

We have a third copy of Z_6 with cx and b^3x^2 as the elements of order 6, bx^2 and c^3x of order 3, and sharing the same Z_2 as above. Now we are only left to check a^2x and a^2x^2.

$(a^2x)^2 = a^2xa^2x = a^2a^2xx = x^2$ $(a^2x)^3 = x^2a^2x = a^2$ $(a^2x)^4 = x^2x^2 = x$
$(a^2x)^5 = xa^2x = a^2x^2$ $(a^2x)^6 = a^2a^2 = 1$

So the last two elements determine another Z_6 but this time it shares the preceding Z_2 and the Z_3 generated by x. Below is the lattice diagram that we have determined by our preceding calculations. It is a group of order 24 but unlike any of the groups of order 24 that we have had before. It is possible there are subgroups of order 12 and that will be left for you to investigate.

The name for this group is **SL(2,3)**; the SL stands for "special linear". The elements of this group are 2 x 2 matries whose entries come from Z_3, and whose determinant equals 1 (computed mod 3, of course). Eventually we will study this family of groups in much more detail.

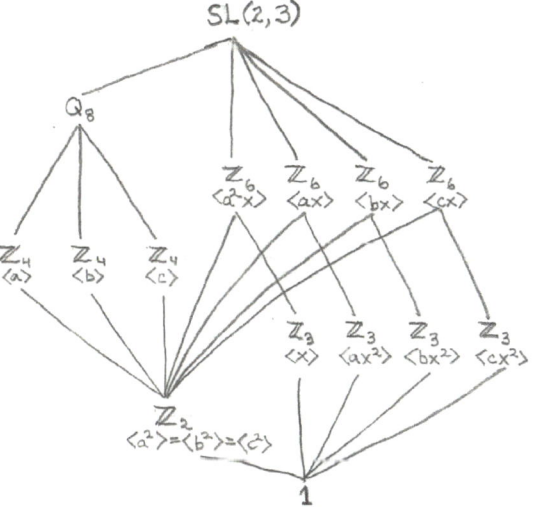

Exercise 63: Referring to lesson 8, find an extension of Z_{12} in which we can extend ϕ_1 to an inner automorphism of the result. Find another extension of Z_{12} in which we can extend ϕ_2 to an inner automorphism.

Exercise 64: Referring to lesson 9, find an extension of D_4 in which we can extend ϕ_2 to an inner automorphism of the result. Do the same for ϕ_3 and ϕ_4.

Exercise 65: Are there subgroups of order 12 in **SL(2,3)**? Derive a presentation for **SL(2,3)** from the presentation for Q_8.

Lesson 18: The Extremes – Simple Groups and Hamiltonian Groups

The normal subgroups of the group G give us an informal measure of how nearly commutative G is. For Abelian groups, every subgroup is normal and Abelian groups are entirely commutative. But if G is non-Abelian the existence of a normal subgroup is quite significant. Here we will consider the two most extreme possibilities: a non-Abelian group which has no non-trivial normal subgroups (the trivial subgroup, and the improper subgroup, are always normal regardless), and a non-Abelian group in which every subgroup is normal. Let's begin with the more important of the two extremes.

Definition 44: A group G which has no non-trivial normal subgroup is called a **simple group**.

There are simple Abelian groups. Any cyclic group of prime order has no non-trivial subgroups at all, and by default, therefore, it has no non-trivial normal subgroups. Thus $\mathbf{Z_p}$ is a whole family of simple groups, and "simple" is the right word to describe them. They are so simple that there is scarcely anything to say about them (at least in group theory; there is additional structure that will be taken into consideration eventually). Among Abelian groups, the $\mathbf{Z_p}$ are the only simple groups.

In a non-Abelian context, the more nearly commutative the group is the easier it is to understand. Hence we would expect that a simple non-Abelian group would be the most complicated non-Abelian groups. That turns out to be the case, making *simple* non-Abelian groups the most inappropriately named object in all of mathematics. In a simple non-Abelian group nearly every vestige of commutativity is absent; *nearly* every vestige but not quite every vestige. After all, every non-Abelian group has cyclic subgroups generated by each of its elements, and each of those cyclic subgroups are internally commutative. It is a question of how far commutativity can be extended beyond those cyclic subgroups. That no group can be purely non-commutative is one indication that commutativity has some indispensable role in algebra.

It turns out that every group can be realized as an extension of some simple group. Thus, if we had a complete list of all the simple non-Abelian groups, we would, at least formally, have a list of every non-Abelian group. The attempt to compile such a complete list of all the simple non-Abelian groups occupied a large proportion of the research in group theory for a century and with the combined efforts of dozens of mathematicians, the list was finally completed in 1980. But every answer leads to more and deeper mysteries than it solves. Finding a list of all these non-Abelian simple groups did not bring group theory to a sense of completion. The unanswered question are different now, but more puzzling than ever.

For the present moment, however, our focus is to introduce the most accessible family of simple non-Abelian groups. In fact it is a family of non-Abelian groups we have already encountered before.

Theorem 74: If $n > 4$ then $\mathbf{A_n}$ is a simple group.

Proof: The following proof is not very elegant. It uses essentially the same trick repeatedly through six separate cases. The first step is to notice that, by Corollary 35A from part 1, $\mathbf{A_n}$ is generated by all the 3-cycles from a set of n letters; it consists in the 3-cycles and all their products. We will prove this theorem by assuming $\mathbf{A_n}$ has a normal subgroup N and showing that N must contain all the 3-cycles and hence must equal $\mathbf{A_n}$. Now if N is normal and includes an element σ, then N must also include all the conjugates of σ as well. So if N includes a 3-cycle then N would contain all the 3-cycles, by theorem 72, and thus would contain all their products and hence would equal $\mathbf{A_n}$. We will call this **case 1**.

Case 2: So we begin with an arbitrary element σ of N and write it as the product of disjoint cycles. In this factorization, suppose σ contains a cycle of length $m > 3$, as well as other cycles of arbitrary length; say

$$\sigma = (a_1, a_2, \ldots, a_m)(\text{ possibly other cycles})$$

Take $\tau = (a_{m-2}, a_m, a_{m-1})$, which is necessarily an element of $\mathbf{A_n}$. Then it is easy to check that

$$\tau^{-1}\sigma\tau = (a_1, a_2, \ldots, a_{m-3}, a_{m-1}, a_m, a_{m-2})(\text{possibly other cycles}),$$
and since N is normal, this must be an element of N. Then it is easy to check that
$$(\tau^{-1}\sigma\tau)\sigma^{-1} = (a_{m-3}, a_m, a_{m-2}) \in N,$$
where all the other possible cycles have cancelled out. Hence N includes a 3-cycle and is therefore equal to A_n.

Case 3: Thus we may assume N has no element that contains cycles longer than 3. When n = 5 this means that the element of N may contain at most one 3-cycle and the rest 2-cycles. So let's temporarily assume n > 5 and let σ be an element of N which is the product of two or more disjoint 3-cycles:
$$\sigma = (a_1, a_2, a_3)(b_1, b_2, b_3)(\text{ possibly other cycles }).$$
This time take $\tau = (a_3, b_2, b_1)$. Then it is easy to check that
$$\tau^{-1}\sigma\tau = (a_1, a_2, b_1)(b_2, a_3, b_3)(\text{ possibly other cycles })$$
$$\text{and } (\tau^{-1}\sigma\tau)\sigma^{-1} = (a_2, b_2, a_3, b_1, b_3) \in N.$$
Thus this case reduces to the previous case.

Case 4: We may now suppose that every element of N has at most one 3-cycle and a series of 2-cycles when factored into disjoint cycles. This case includes what we had set aside in the previous case when we temporarily assumed n > 5. Let the 3-cycle contained in σ be (x, y, z). Then we have $\sigma^2 = (x, z, y) \in N$ because all the transpositions cancel out. Thus N contains a 3-cycle and so $N = A_n$.

Case 5: The last possibility is that no element σ contains a cycle longer than a transposition when factored into disjoint cycles. We need not even require that these transpositions be mutually disjoint, but we do know that there must be an even number of these cycles since A_n contains only even permutations. Let's assume
$$\sigma = (x, y)(z, u)(\text{ possible other transpositions })(a, b)(c, d)$$
where some of the transpositions may have letters in common. This time take $\tau = (y, a)(b, c)$ which is an even permutation and so is an element of A_n. Then
$$\tau^{-1}\sigma\tau = (x, a)(z, u)\ldots(y, c)(b, d)$$
$$\text{and } (\tau^{-1}\sigma\tau)\sigma^{-1} = (x, c, b)(y, a, d).$$
This is back to case 3, which required n > 5. So $N = A_n$.

Case 6: We are left with only one possibility: N includes an element that consists in exactly two 2-cycles, say $\sigma = (a, b)(c, d) \in N$. Since we have at least five letters, σ must fix at least that letter, say e, and possibly more. Take $\tau = (a, e, b) \in A_n$. Then $\tau^{-1}\sigma\tau \in N$ and $(\tau^{-1}\sigma\tau)\sigma^{-1} = (a, e, b) \in N$. This takes us back to case 1.

Therefore it is always true that if $N \triangleleft A_n$, N is either trivial or else $N = A_n$.

Corollary 74a: If n > 4 the only proper normal subgroup of S_n is A_n.
Proof: this is left as an exercise.

As it turns out, A_5, of order 60, is the smallest non-Abelian simple group. Including Abelian groups, Z_p and A_n for n > 4 make two infinite families of simple groups. There are 16 other such infinite families of simple non-Abelian groups, some of which we will eventually study. In addition to these infinite families, there are 26 simple non-Abelian groups which do not fit into such an infinite family. They are called *sporadic* simple groups. We will learn about a few of these in part 4 of this series.

At the other end of the spectrum there are non-Abelian groups for which every subgroup is normal. They deserve a special name:

Definition 45: A non-Abelian group in which every subgroup is normal is called a **Hamiltonian group**.

These are the most nearly commutative non-Abelian groups, as far as normality is an indication, and are interesting for that reason alone, if no other. You might wonder if any Hamiltonian groups exist at all, but in fact we have already encountered one, namely the group Q_8. Hamiltonian groups are rarer than simple groups and listing them is not so arduous. We do not yet have the tools to do it, however, and we will return to them in part 3. They do provide us with an interesting counter-example, though, of a direct product which has a non-normal subgroup though both of its factors have only normal subgroups.

Exercise 66: Prove Corollary 74a.

Exercise 67: Verify that Q_8, $Q_8 \times Z_2$, and $Q_8 \times V$ are Hamiltonian groups.

Exercise 68: Show that $Q_8 \times Z_4$ has a subgroup that is not normal.

> William Rowan Hamilton was born in Dublin, Ireland, in 1805, the fourth of nine children. He showed an aptitude for both languages and mathematics very early. By the time he was 12 he had studied the classical and modern European languages and in addition had learned Hebrew, Arabic, Persian, Hindustani, Sanskrit, Marathi and Malay. In 1823 he entered Trinity College where he studied both classics and mathematics and earned the highest possible ranks in both. While still an undergraduate, in 1827, he became the Royal Astronomer of Ireland and lived the rest of his life at Dunsink Observatory. That same year he published a paper which firmly established the wave theory of light as the correct viewpoint. In 1835 he reformulated physics in the form now known as Hamiltonian mechanics, a form which proved most suitable for the development and electromagnetic theory and eventually quantum mechanics. His formulation is still at the center of modern physics. In 1843 he discovered the quaternions as an extension of the complex numbers, and the group of quaternions as a subscript. The quaternions were the first algebraic system to deliberately abandon the commutative property and helped set algebra free to explore other axiomatic systems. The quaternions are still used as a convenient way to express rotations in a three dimensional space. Early in his life Hamilton tried his hand at poetry, but his friend Wordsworth recognized his lack of talent and discouraged him. It is believed that Charles Dodgson (Lewis Carroll) was drawing a caricature of Hamilton in the character of the Mad Hatter in *Alice's Adventures in Wonderland*. Hamilton died in 1865 after a severe spell of gout.

Lesson 19: The Extremes – Elementary Groups

Just as it is worthwhile knowing all the simple groups, and to a lesser extent all the Hamiltonian groups, it is also worthwhile knowing all those groups which have no characteristic subgroups. A group may have a normal subgroup which is not characteristic, so groups which have no characteristic subgroups may still have plenty of normal subgroups; but if a group has no normal subgroups then naturally it can have no characteristic subgroups.

Definition 46: Let G be a group which has no proper characteristic subgroups. We will call such a group an **elementary group**.

A group may have a normal subgroup which is not characteristic, so elementary groups need not be simple. But if a group is simple then naturally it is also elementary. So in making a list of all the elementary groups we can begin by inserting all the simple groups. We need one more definition to continue.

Definition 47: Let G be any group. A **minimal normal subgroup** of G is a proper normal subgroup of G which contains no proper subgroup that is normal in G.

Theorem 75: A group G is elementary iff it is the direct product of a finite number of copies of a single simple group.

Proof: ⇒) We assume G is elementary. If G has no normal subgroups, then we are done. So let N be a minimal normal subgroup. Let β be any automorphism of G. Then β(N) is also a minimal normal subgroup. Form the subgroup generated by all the minimal normal subgroups which are images of N: C = < β(N) | for β ∈ **Aut**(G) >. Clearly C is mapped to itself by any automorphism of G, and therefore C ≪ G. Therefore C = G since G is elementary.

Now any two images of N must intersect in the identity since each one is a minimal normal subgroup of G and a non-trivial intersection would be normal if it existed. Then, by theorem 62, G is isomorphic to the direct product of some of these images of N (it is left to the student to extend theorem 62 to more than two normal subgroups). Without loss of generality, we may assume that N is one of these factor groups. Then N has no normal subgroup; if it did, by theorem 55, that normal subgroup would be normal in G, contradicting the minimality of N. Hence N is itself a simple group and G is isomorphic to the direct product of identical simple groups.

⇐) Now assume G is the direct product of a finite number of identical copies of a particular simple group S: $G \simeq S_1 \times S_2 \times \ldots \times S_n$ where each $S_i \simeq S$. Let H be some proper characteristic subgroup of G. Then for some i, $H \cap S_i \neq 1$. Further $H \cap S_i \triangleleft G$ by theorem 68. Since S_i is simple, it must be the case that $H \cap S_i = S_i$ so $S_i \leq H$. Let S_j be any other of the factors. Then there is an automorphism β of G defined by

$$\beta(x_1, \ldots, x_i, \ldots, x_j, \ldots, x_n) = \beta(x_1, \ldots, x_j, \ldots, x_i, \ldots, x_n)$$

Hence β maps S_i onto S_j. But β preserves H since H is characteristic. Therefore H must contain all of the factors S_k, and therefore H = G. Hence G contains no proper characteristic subgroups and is an elementary group.

Thus we know all of the elementary Abelian groups. For a particular prime, p, the elementary Abelian groups are just the groups $\mathbf{Z_p} \times \mathbf{Z_p} \times \ldots \times \mathbf{Z_p}$ for a finite number of factors. We will denote the elementary Abelian group of order p^n by $\mathbf{E_{p^n}}$. The Klein 4-group is one of these elementary groups and so it might be denoted by $\mathbf{E_4}$ as well as by **V**. The elementary non-Abelian groups we do not know because we do not know all the simple non-Abelian groups. But we do know that $\mathbf{A_n} \times \mathbf{A_n} \times \ldots \times \mathbf{A_n}$ will be elementary when n is greater than 4. Thus the smallest non-simple elementary non-Abelian group we know at this point has order 3600.

Theorem 76: Let G be any group. Then H is a minimal normal subgroup of G iff H is an elementary subgroup of G.
Proof: use theorem 66.

We can follow the same chain of inquiry we pursued in lesson 17: characterize all the groups for which every proper subgroup is characteristic. This is a more difficult to answer than it was for the case of Hamiltonian groups. But at least we can easily see that the cyclic groups all fit in this class.

What theorem 76 tells us is that every group will have one or more minimal normal subgroups and that those will be elementary. This means that if a group is large enough, we can expect to find subgroups which are isomorphic to the direct product of some prime cyclic groups; it may be that the minimal normal subgroups are the prime cyclic subgroups but we know we should look for their direct products. This warns us of the possible presence of subgroups we might not have looked for otherwise.

The next theorem is a basic property of the minimal normal subgroups which is sometimes useful.

Theorem 77: Let K be a minimal normal subgroup of G and let L ◁ G. Then either K ∨ L ≃ K × L or else K ≤ L.
Proof: Suppose K is not a subgroup of L. Then K ∩ L ◁ G by theorem 63a. Since K is minimal, however, this means that K ∩ L = **1**. Since K and L are both normal in K ∨ L, theorem 62 tells us that K ∨ L ≃ K × L.

Corollary 77a: Let K be a minimal normal subgroup of G and let L ◁ G. If K is not a subgroup of L, then the elements of K commute with the elements of L.
Proof: this follows from theorem 60.

We will take this material one more step with the introduction of a new term.

Definition 48: The subgroup of G which is generated by all the minimal subgroups of G is called the **socle** of G and is denoted by **soc**(G).

Since the socle is the join of the minimal normal subgroups, any two of which have a trivial intersection, it is clear that **soc**(G) is the direct product of the minimal normal subgroups. It is not elementary unless there is only one minimal normal subgroup because the different minimal normal subgroups will be the direct products of different simple subgroups. In the case of infinite groups, non-trivial minimal normal subgroups may not exist, but we will discuss that in part 5.

Theorem 78: For any group G, **soc**(G) ≪ G.
Proof: is an easy exercise.

We will consider, as a few simple examples, all the groups we know of order 8. The Abelian ones are the easiest since all their subgroups are normal. In Z_8 the only minimal normal subgroup is Z_2, and hence **soc**(Z_8) = Z_2. In $Z_4 \times Z_2$, there are three minimal normal subgroups, the three copies of Z_2; together they generate the **V** subgroup, so we have **soc**($Z_4 \times Z_2$) = **V**. In $Z_2 \times Z_2 \times Z_2$ the minimal normal subgroups are the seven copies of Z_2 which together generate the whole group. Therefore **soc**($Z_2 \times Z_2 \times Z_2$) = $Z_2 \times Z_2 \times Z_2$.

Considering the non-Abelian groups, the quaternion group Q_8 has exclusively normal subgroups. Its only minimal normal subgroup is the sole copy of Z_2, and therefore **soc**(Q_8) = Z_2. Finally consider D_4. We know the normal subgroups of D_4 are the two copies of **V**, the copy of Z_4 and the Z_2 they all have in common. Hence that particular copy of Z_2, which is generated by the 180° rotation, is the only minimal normal subgroup; the other copies of Z_2 are not normal. Hence, **soc**(D_4) = Z_2.

Exercise 69: Prove theorem 76.
Exercise 70: Let G be any cyclic group. Show that every subgroup of G is characteristic in G.
Exercise 71: Find **soc**(D_6), **soc**(Q_{16}), **soc**(A_4) and **soc**($Z_3 \times D_4$). Refer to lesson 4 for the last one.
Exercise 72: Prove theorem 78.

Lesson 20: Quotient Groups

We come now to the most fruitful idea that grows out of the concept of the normal subgroup. It arises from the unique property of normal subgroups that their corresponding left and right cosets are always equal. For normal subgroups, we can drop the left and right distinction and just speak of cosets without qualification. So let G be a group and let H ◁ G. Now consider the following simple steps:
$$aH \cdot bH = a(Hb)H = a(bH)H = abHH = abH$$
It should be clear that HH = H; it just says that an element in H times and element in H equals an element in H. What we have here is nothing less than a coherent way to define the multiplication of cosets. It will perhaps be clearer if we modify the notation slightly. Given representatives of a coset, say a and b, use **a** = aH and **b** = bH so that we have **ab** = abH. Then aH·bH = abH takes on the more obvious appearance **a·b** = **ab**.

This way of defining coset multiplication depends critically on being able to switch between left and right cosets at will. There is no way to define a straight forward operation on the cosets of a non-normal subgroup. The pivotal theorem for this chain of ideas is the following.

Theorem 79: Let G be a group and H ◁ G. Then the set of all the cosets of H form a group under the operation of coset multiplication.

Proof: First, is coset multiplication a well-defined operation on the collection of cosets? In other words, if we choose different representatives for the cosets involved in a product, will we get a different answer? Let a' ∈ aH and b' ∈ bH; is a'Hb'H = aHbH? Since a' ∈ aH, a' = ah_1; similarly b' = bh_2. Then a'Hb'H = ah_1H bh_2H = aHbH. Thus coset multiplication is a well-defined on cosets.

Coset multiplication is associative because the operation of the group G is associative.

The coset H itself serves as the identity since aHH = aH.

The inverse of the coset aH is a^{-1}H since aHa^{-1}H = aa^{-1}H = H.

Finally the product of any two cosets is clearly a coset. Therefore the set of all cosets of a normal subgroup form a group under coset multiplication.

This leads us to coin a new term:

Definition 49: Let G be a group and H ◁ G. The group that is formed by the collection of the cosets of H in G under coset multiplication is called the **quotient group of G by H** and is denoted by G/H.

It is important to keep in mind that the elements of a quotient group are in fact whole sets of elements in the original group. You must learn to think on two different levels when it comes to the elements you are working with. Let's look at two examples of quotient groups to see in detail how the process of forming the quotient works out. We will do this by taking the Cayley table and explicitly showing how a normal subgroup splits the group into blocks that work coherently with each other to form a new group. First I will give the Cayley table of the group as it is usually given; then I will rearrange the elements into the cosets so they exhibit clearly the multiplication in the quotient.

First let's consider Z_{12}/Z_4. Since we are beginning with an Abelian group we have no worry that the subgroup might not be normal. Z_{12} is usually given in additive notation, and so our quotient will use the operation of coset addition; instead of writing aH·bH = abH we write (a + H) + (b + H) = a + b + H. First we list the elements in the quotient, which are the cosets of Z_4:

$$Z_4 = \{0, 3, 6, 9\} \qquad 1 + Z_4 = \{1, 4, 7, 10\} \qquad 2 + Z_4 = \{2, 5, 8, 11\}$$

In the right hand Cayley table you can see clearly how the blocks multiply each other to give other blocks in the same collection of blocks. Once the elements are rearranged and we view the table in terms of the blocks we see that we have the Cayley table for Z_3. Hence what we have illustrated here is $Z_{12}/Z_4 \simeq Z_3$. Usually this is read "Z_{12} mod Z_4 is Z_3". For cyclic groups, quotients look like division, but not for non-cyclic groups.

Now let's consider a non-Abelian example, **D₆**, and choose the normal subgroup of rotations by $0°$, $120°$, and $240°$. We have not shown that this subgroup, which is a copy of **Z₃**, is actually normal, but it is easy to see; and the fact that the cosets form the proper blocks for multiplying is also proof. Enumerating the cosets we have

$\mathbf{Z_3} = \{\rho_0, \rho_2, \rho_4\}$ $\quad\quad \rho_1\mathbf{Z_3} = \{\rho_1, \rho_3, \rho_5\}$ $\quad\quad \delta_1\mathbf{Z_3} = \{\delta_1, \delta_2, \delta_3\}$ $\quad\quad \mu_1\mathbf{Z_3} = \{\mu_1, \mu_3, \mu_2\}$

Working the same rearrangement in the Cayley table for **D₆** gives

This time, looking only at the cosets, the Cayley table is plainly that of **V**. Thus we have shown that $\mathbf{D_6}/\mathbf{Z_3} \simeq \mathbf{V}$; that is, **D₆** mod **Z₃** is **V**. Now several facts about quotient groups are easy to prove:

Theorem 80: $|G/H| = |G| / |H|$.

Proof: is left to the exercises.

....

As was true for direct products, there is a distributive effect when we mod out a subgroup H. H will be modded out of each subgroup of G as well, in so far as H intersects the subgroup. We will exploit this a great deal in future work, especially in part 3. One question that begs to be addressed is: what is the connection between quotients and direct products? Are they opposites? The answer is: they are partial opposites.

Theorem 81: If $G \simeq H \times K$ then $H \simeq G/K$ and $K \simeq G/H$.
Proof: is left to the exercises.

However the converse of theorem 81 is not true. In the example above we found $D_6/Z_3 \simeq V$ but D_6 is not isomorphic to $Z_3 \times V$.

Theorem 82: If G is Abelian and H is any subgroup of G, then G/H is Abelian.
Proof: is left to the exercises.

We are now able to define outer automorphisms:

Definition 50: An **outer automorphism** of the group G is an element of **Aut**(G)/**Inn**(G).

An outer automorphism is not an individual automorphism, but a whole coset of automorphisms. Outer automorphisms that only differ by a factor of an inner automorphism are considered equivalent to each other.

Exercise 73: Prove theorem 80.
Exercise 74: Prove theorem 81.
Exercise 75: Prove theorem 82.
Exercise 76: Compute the quotient Q_{16}/Z_4 where the Z_4 we are modding out is $<x^2>$. (The lattice diagram is given in lesson 30 of part 1.) Though you know what the quotient must be simply from the orders of the group and subgroup involved, construct the quotient in detail by way of the Cayley table as done above.
Exercise 77: Let $A \triangleleft G$ and $B \triangleleft H$. Show that $(G \times H)/(A \times B) \simeq (G/A) \times (H/B)$.
Exercise 78: Let G be Abelian and let $D = \{(g, g) \in G \times G \}$. Show that $(G \times G)/D \simeq G$. Give a counter-example to show that this fails if G is non-Abelian.
Exercise 79: Let $A \triangleleft G$, $B \triangleleft G$, and $G = AB$. Show that $G/(A \cap B) \simeq (G/A) \times (G/B)$.
Exercise 80: Show that if G/A and G/B are both Abelian, then $G/(A \cap B)$ is Abelian.
Exercise 81: Using the results of lesson 9, find all the distinct outer automorphisms of Q_8 and of D_4. Use what you know of Hamiltonian groups.

Lesson 21: Homomorphisms

We now come to the primary goal of this chapter of modern algebra. Most students of algebra encounter the material in this lesson and the next four lessons as the first great milestone of algebraic insight. Everything so far in our study has been leading to this as its first great goal. And everything that we will study in the future development of algebra will flow out from this point as its source.

I have divided this material into a cluster of eight theorems, theorems so important that they have names as well as numbers. Usually these results are presented in fewer pieces than eight, sometimes as few as three, but I hope that by separating the results into more parts they will be more easily assimilated. I also hope it will be easier to focus on the details if we take the results somewhat slowly. These theorems provide necessary tools for the study of large groups which are too large to know in detail. When a group has hundreds or thousands of elements – and we will want to look at groups much larger than thousands of elements – we will need to understand more of the general behavior of subgroups, especially the normal ones.

These theorems are concerned with the rather intricate relationships that exist between subgroups that are normal in other subgroups, and the relationships that exist between the quotient groups of those pairs. We will examine these intricacies in some detail, even when the knowledge itself is not necessary to our purposes for quite a long time. The goal by the end of this part of Modern Algebra is to have some well developed facility with thinking about quotient groups.

In lesson 1 of this part of Modern Algebra, in my definition of the isomorphism, I required that it be a bijection; in particular, it had to map a group G *onto* another group H. We will now introduce a more general class of algebra preserving functions which include isomorphisms as a special type.

Definition 51: Let G and H be groups. Let $\phi: G \to H$ be a function such that $\phi(xy) = \phi(x)\phi(y)$ for all pairs $x, y \in G$. ϕ is called a **homomorphism** from G into H. If $\phi(x) = 1$ for all $x \in G$, then ϕ is called the **trivial homomorphism**.

Definition 52: Let G and H be groups and $\phi: G \to H$ be an injective homomorphism. Then ϕ is called an **inclusion map** and we denote it by $\phi: G \hookrightarrow H$.

As an example of an inclusion map, suppose $G \simeq H \times K$. Then the map $\phi: H \to G$ as defined by $\phi(h) = (h, 1)$ is an inclusion map. Sometimes inclusion maps are called *insertions*, sometimes they are called *imbeddings* and sometimes they are called *injections*, but I will try to consistently refer to them as inclusions. Inclusions are almost isomorphisms. Some texts prefer to not require that isomorphisms be surjective, and for them an inclusion is an isomorphism. Certainly it is true that an inclusion $\phi: G \hookrightarrow H$ establishes an isomorphism (my definition) between G and $\phi(G) \leq H$.

Though I have said it already, I will say it again: be sure to note that every isomorphism is also a homomorphism by definition, and that a homomorphism can map several elements of G to the same element of H. Being a function, of course, a homomorphism is *not permitted* to map an element of G to several elements of H. Our next theorem summarizes the properties of all homomorphisms.

Theorem 83: Let $\langle G, +\rangle$ and $\langle H, \cdot\rangle$ be groups and $\phi: G \to H$ be a homomorphism. Then we have

   a) if 0 is the identity of G and 1 is the identity of H, then $\phi(0) = 1$.
   b) for any $g \in G$, $\phi(-g) = \phi(g)^{-1}$.
   c) for any $g \in G$, $\phi(ng) = \phi(g)^n$.
   d) if g has order n in G, then the order of $\phi(g)$ divides n.
   e) if $K \leq G$, then $\phi(K) \leq H$.
   f) if $\phi: G \to H$ and $\psi: H \to K$ are homomorphisms, then the composition $\psi \circ \phi: G \to K$ is a homomorphism.

Proof: most of this follows immediately from the definition of homomorphism and the properties we have already proven for isomorphisms in theorem 37. You should be able to prove these easily as exercises, but pay special attention to part **d)** which is the only one that is significantly different.

In theorem 37 there was a part g. It stated that the inverse of an isomorphism is an isomorphism. But the inverse of a homomorphism may not even be a function, much less a homomorphism. Even when ϕ^{-1} is not a homomorphism, we will make use for it later.

The most significant element of a group is the identity element, and it is even more significant to the discussion of homomorphisms than it was for isomorphisms. A homomorphism ϕ may map several elements of G to the identity of H, and these elements are so important they get a special name.

Definition 53: Let $\phi : G \to H$ be a homomorphism. We will call the set $\{g \in G \mid \phi(g) = 1\}$ the **kernal** of ϕ and denote it by **ker**(ϕ).

The kernal of a homomorphism plays a primary role in the details of how the homomorphism maps the rest of the group, as this next theorem shows.

Theorem 84: Let $\phi: G \to H$ be a homomorphism. Then **ker**(ϕ) $\triangleleft$ G.

Proof: To simplify our writing, we will usually let K = **ker**(ϕ). First we need to show that K is a subgroup. Clearly $1 \in K$. If $k \in K$, then $\phi(k) = 1$, by definition and by theorem 83b, $\phi(k^{-1}) = 1^{-1} = 1$; K contains the inverses of each of its elements. Last if $k_1, k_2 \in K$, then $\phi(k_1 k_2) = \phi(k_1)\phi(k_2) = 1 \cdot 1 = 1$, so $k_1 k_2 \in K$. Thus $K \leq G$.

To show that K is normal, we need to show that K includes the conjugates of each of its elements. So let $k \in K$ be an arbitrary element and $g \in G$ also be arbitrary. Then
$$\phi(g^{-1}kg) = \phi(g^{-1})\phi(k)\phi(g) = \phi(g)^{-1}\phi(k)\phi(g) = \phi(g)^{-1} \cdot 1 \cdot \phi(g) = \phi(g)^{-1}\phi(g) = 1$$
Hence $g^{-1}kg \in \mathbf{ker}(\phi)$ and therefore $\mathbf{ker}(\phi) \triangleleft G$.

Sometimes we use $\phi^{-1}(1)$ as an alternative notation for **ker**(ϕ). This is an abuse of notation since ϕ^{-1} may not be a function. It causes no difficulty, though, as long as we keep our wits about us. This leads to further terminology, a new term which has become common recently.

Definition 54: Let $\phi: G \to H$ be a homomorphism. For any $h \in H$, the set $\phi^{-1}(h) = \{g \in G \mid \phi(g) = h\}$ is called the **fiber** of ϕ over h.

The next theorem shows how the knowledge of the kernal of the homomorphism ϕ tells us everything we need to know about how ϕ maps G.

Theorem 85: Let $\phi: G \to H$ be a homomorphism, K = **ker**(ϕ), let $h \in H$ be an arbitrary element of H and let j be an arbitrary element of the fiber over h. Then the fiber over h equals the coset of K represented by j: $\phi^{-1}(h) = jK$.

Proof: Let j be any element of the fiber over h, so $\phi(j) = h$. First we will show that the fiber is contained in the coset jK. Note that $\phi(j^{-1}) = \phi(j)^{-1} = h^{-1}$. We need to show that if y is any other element of the fiber over h then also $y \in jK$. Since y is in the fiber over h, $\phi(y) = h$. Now consider
$$\phi(j^{-1}y) = \phi(j^{-1})\phi(y) = h^{-1}h = 1.$$
Hence $j^{-1}y \in K$. Therefore $y \in jK$ and $\phi^{-1}(h) \subseteq jK$.

The reverse containment is easier. Let j be any element of the fiber over h, and let y be any element of the coset jK. Then y = jk for some element $k \in K$. Then
$$\phi(y) = \phi(jk) = \phi(j)\phi(k) = h \cdot 1 = h$$
Hence y is in the fiber over H. Therefore $jK \subseteq \phi^{-1}(h)$. Hence we have shown $jK = \phi^{-1}(h)$.

Now we can think of the elements of a quotient group in either of two ways, either as cosets of a normal subgroup, or as the fibers of elements from a homomorphism. We have established a very tight connection between homomorphisms and the normal subgroups that are their kernals. The theorems in the next lesson will strengthen this connection considerably.

One more important theorem to complete this lesson. We showed that the maps λ_g and ρ_g were not isomorphisms, but there is another way of interpreting them. Each of λ_g and ρ_g maps G onto G and thus can be thought of as permutations of the elements of G; that is, λ_g and $\rho_g \in S_n$ where n = |G|. So:

Theorem 86 (Cayley's Theorem): The map $\psi: G \to S_n$ where n = |G| defined by $\psi(g) = \lambda_g$ is a homomorphism.

Proof: is left as an exercise.

Thus every group is isomorphic to some subgroup of some S_n. We can think of any group as a group of permutations. Just as cyclic groups could be thought of as the atomic groups, the symmetric groups can be thought of as the universal groups. Theorem 86 is named for the same Cayley who gave us Cayley tables and whose biographical sketch is in part 1.

Exercise 82: Prove all the parts of theorem 83 but prove them without appealing to theorem 37.

Exercise 83: Let $\phi: G \to H$ be a homomorphism. Show that ϕ is an isomorphism iff **ker**$(\phi) = 1$.

Exercise 84: Show that the map $\phi: G \to G$ defined by $\phi(g) = g^{-1}$ is a homomorphism (in fact an automorphism) iff G is Abelian.

Exercise 85: Show that the map $\phi: G \to G$ defined by $\phi(g) = g^2$ is a homomorphism iff G is Abelian. When is it an automorphism?

Exercise 86: Prove theorem 86. What about the map $\chi(g) = \rho_g$? (maps like χ are called *anti-isomorphisms*).

Lesson 22: The First Isomorphism Theorem

The two theorems in this section are usually put together into a single larger theorem. It is sometimes called the Fundamental Theorem of Homomorphisms and sometimes the First Isomorphism Theorem. I have elected to split the theorem into two parts that are converses of each other and use each name for its appropriate half.

Theorem 87: (The Fundamental Theorem of Homomorphisms) Let G be any group and K ◁ G be any non-trivial normal subgroup of G. Then there is a unique homomorphism γ_K: G → G/K such that **ker**(γ_K) = K.

Proof: We will first define a candidate homomorphism in the most straight-forward way: $\gamma_K(g)$ = gK. γ_K is clearly a surjection, though that is not required to make it a homomorphism. Does it preserve products? On the one hand $\gamma_K(gh)$ = ghK. On the other hand $\gamma_K(g)\gamma_K(h)$ = gKhK = ghK by definition of coset multiplication. Hence γ_K is a homomorphism.

For uniqueness, let ϕ: G → G/K be another homomorphism. Now ker(ϕ) = {x ∈ G | ϕ(x) = **1**} by definition, where the identity of G/K is the coset K. Thus ker(ϕ) is the set of all elements, x ∈ G such that ϕ(x) ∈ K. Now let g be an arbitrary element of G, and let ϕ(g) = hK ∈ G/K. By theorem 85, the fiber over hK under the homomorphism ϕ equals the coset hK in G, and by definition g is also in that fiber. Thus gK = hK. Hence we see that ϕ(g) = hK = gK = $\gamma_K(g)$. So γ_K and ϕ agree on every element of G, and are therefore the same function. Hence γ_K is unique.

Definition 55: The unique homomorphism γ_K given by theorem 87 is called the **canonical homomorphism** from G to G/K. γ_K is also sometimes called the **natural homomorphism**.

Thus we now know that every normal subgroup completely determines a homomorphism. There are always at least as many homomorphism that can be defined on a group as there are normal subgroups. We now consider the converse.

Theorem 88: (The First Isomorphism Theorem) Let G be any group and ϕ: G → H be any surjective homomorphism that is not an isomorphism. Let K = **ker**(ϕ). Then there exists a unique isomorphism ψ: G/K ≃ H.

Proof: The isomorphism ψ must map cosets of K in G to elements of H. Let ψ: G/K → H be the candidate isomorphism defined by ψ(gK) = ϕ(g). In this case, we must check if ψ is well-defined; in other words, if we choose a different representative of the coset gK, will the value of ψ remain unchanged? To check let j ∈ gK. Then j = gk for some k ∈ K. Then
$$\psi(jK) = \psi(gkK) = \psi(gK)$$
Hence ψ is well-defined. Since ϕ is a surjection, ψ is also surjective. Now if ψ(xK) = ψ(yK) then ϕ(x) = ϕ(y). Thus the fiber over x under ϕ equals the fiber over y under ϕ; that is, xK = yK. Hence ψ is injective. Finally we show that ψ preserves products:
$$\psi(xyK) = \phi(xy) = \phi(x)\phi(y) = \psi(xK)\psi(yK).$$
The proof of uniqueness for the isomorphism is left as an exercise.

Thus every distinct homomorphism determines a distinct normal subgroup given by its kernal, and there is a one-to-one correspondence between the possible homomorphisms of a group and its proper normal subgroups. This theorem allows us to introduce another kind of diagram which becomes increasingly important as algebra develops. It is called a **commutative diagram**. We first write down the various groups (or other algebraic structures we have not studied yet) along with the functions that map one to another. The resulting diagram of all the functions is said to be *commutative* if whenever we follow consecutive functions to get from one group to another, we always get the same result whichever path we take. Theorem 87 gives us a particularly simple commutative diagram involving only three groups. Future commutative diagrams may involve a large number of algebraic structures and become quite complex. The value of a commutative diagram lies in displaying which compositions of functions are equal. Here I show the commutative diagram that displays both the Fundamental Theorem of Homomorphisms and the First Isomorphism Theorem.

Let's examine the special case of direct products more closely. Beginning with $G \simeq H \times K$, then we have already noted that both factor groups are normal in G and the result of modding out either factor group is the other factor group, $G/H \simeq K$. Then the Fundamental Theorem of Homomorphisms tells us that there is a canonical homomorphism $\gamma_K: G \to H$. In this case the canonical homomorphism assumes a particularly simple form: $\gamma_K((h, k)) = h$. γ_K maps each ordered pair to its first coordinate. In the same way γ_H maps each ordered pair to its second coordinate. We have a special term for the canonical homomorphism in the context of a direct product.

Definition 56: The homomorphism $\pi_H: H \times K \to H$ given by $\pi_H((h, k)) = h$ is called the **projection** of $H \times K$ onto H.

Be careful to notice here that the meaning of the subscript is reversed between canonical homomorphisms and projections. In the canonical homomorphism, the kernal is the subscript; in the projection it is the image of the map that is the subscript. Thus $\gamma_H = \pi_K$ and $\gamma_K = \pi_H$. When we speak of direct products, we will usually refer to the projection map rather than to the canonical homomorphism, but not consistently. Even outside of the context of the direct product we sometimes use the term projection instead the canonical homomorphism. The custom is to use them interchangeably, though a projection map was intended to mean a special case. I will make the effort to only use the term projection to refer to direct products, however.

The direct product, $G \simeq H \times K$, gives rise to the commutative diagram pictured above.

Exercise 87: Find an example of two non-isomorphic groups, G and G', which have normal subgroups $K \triangleleft G$ and $K' \triangleleft G'$ such that $K \simeq K'$ and $G/K \simeq G'/K'$.

Exercise 88: Find an example of a group G with two normal subgroups $K \triangleleft G$ and $K' \triangleleft G'$ such that $K \simeq K'$ but for which G/K is not isomorphic to G/K'.

Exercise 89: Finish the proof of theorem 88.

Exercise 90: Why did we have to assume the kernals were non-trivial in theorems 87 and 88?

Lesson 23: The Correspondence Theorems

Most texts present this material in a single theorem, but I have divided it into two theorems. Together these two theorems allow us to derive the lattice diagram of a quotient group from the lattice diagram of its parent group.

Theorem 89 (The First Correspondence Theorem): Let $N \triangleleft G$ and let $N \leq A$ and $N \leq B$. Then the following are true:
  a) $A \leq B$ iff $A/N \leq B/N$.
  b) if $A \leq B$ then $[B:A] = [B/N:A/N]$.
  c) $A \triangleleft B$ iff $A/N \triangleleft B/N$.

Proof: a) $\Rightarrow$) We begin by assuming that $N \leq A \leq B \leq G$ and $N \triangleleft G$. Since N is normal in G, N is also normal in both A and B so the quotients A/N and B/N are defined. Every coset of N by an element of A is also a coset of N by an element of B since $A \leq B$. Therefore $A/N \leq B/N$.

$\Leftarrow$) Now suppose $N \leq A$, $N \leq B$, $N \triangleleft G$, and $A/N \leq B/N$. Then $aN \in A/N$ implies $aN \in B/N$, which says that aN is a coset in the collection of cosets of N by elements of B. So there exists some element $b \in B$ such that $aN = bN$. Therefore $a = bn$ for some $n \in N$. Since $N \leq B$, $a \in B$. Thus $A \leq B$.

b) For clarity we will now use boldface to denote cosets when they are considered as elements of the quotient, so $\mathbf{a} = aN$ and we will use $\mathbf{A} = A/N$, $\mathbf{B} = B/N$. Let $b_1A, b_2A, \ldots, b_nA$ be a complete collection of distinct cosets of A in B, so that $n = [B:A]$. Then I claim $\mathbf{b_1 A}, \mathbf{b_2 A}, \ldots, \mathbf{b_n A}$ is a complete collection of distinct cosets of $\mathbf{A}$ in $\mathbf{B}$. This list must contain all the cosets of $\mathbf{A}$ in $\mathbf{B}$; to see that there are no duplicates, suppose $\mathbf{b_i A} = \mathbf{b_j A}$. Then $\mathbf{b_i} = \mathbf{b_j a}$ for some $\mathbf{a} \in \mathbf{A}$. But this means that $(b_iN)\mathbf{A} = (b_jN)\mathbf{A}$. Since N is the identity element in $\mathbf{A} = A/N$, $b_i(N\mathbf{A}) = b_j(N\mathbf{A})$ and so $b_i \mathbf{A} = b_j \mathbf{A}$. Therefore $\mathbf{b_1 A}, \mathbf{b_2 A}, \ldots, \mathbf{b_n A}$ is a complete collection of distinct cosets of $\mathbf{A}$ in $\mathbf{B}$, and $[\mathbf{B} : \mathbf{A}] = n$.

c) is left as an exercise.

Hence, taking the quotient of a group preserves the subgroup relationships between all the subgroups that contain the subgroup being modded out. Taking the quotient even preserves the normality of the relationships and the indices of those relationships. But quotients also preserve joins and intersections of subgroups containing the modded out subgroup.

Theorem 90 (The Second Correspondence Theorem): Let $N \triangleleft G$ and let $N \leq A \cap B$. Then the following are true:
  a) $(A \vee B)/N = (A/N) \vee (B/N)$.
  b) $(A \cap B)/N = (A/N) \cap (B/N)$.

Proof: a) We assume $N \triangleleft G$ and $N \leq A, B \leq G$. Since N is normal in G, N is normal in $A \vee B$ and the quotient $(A \vee B)/N$ makes sense. $(A \vee B)/N$ consists in all the cosets of N contained in $A \vee B$, and so it contains all the cosets of N in A and all of the cosets of N in B. Therefore $(A \vee B)/N$ contains A/N and B/N and thus $(A/N) \vee (B/N) \leq (A \vee B)/N$. To show the reverse containment, we begin with $xN \in (A \vee B)/N$. Then x is a product of elements of A and elements of B. By the definition of coset multiplication, these can be factored into cosets:
$$xN = a_1b_1a_2b_2\ldots a_kb_kN = a_1Nb_1Na_2Nb_2N\ldots a_kNb_kN \in (A/N) \vee (B/N)$$
Hence $(A \vee B)/N \leq (A/N) \vee (B/N)$. Therefore the two are equal.

  b) Proof is left as an exercise.

The Second Correspondence Theorem is illustrated in the lattice diagram on the next page. The normality relation between N and G is represented by the double line, and the canonical homomorphism is shown by the dashed red lines.

These two theorems give us a way of deriving a great deal of information about the lattice diagram of a quotient group if we know something about the quotients of its subgroups. It is straightforward to find the quotients of the subgroups containing the normal subgroup. What these theorems help us do is find the quotients of those subgroups which do not contain the normal subgroup.

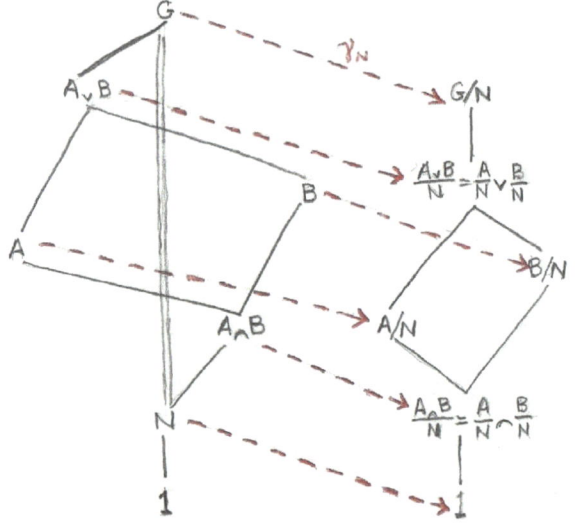

As one example, we can consider $\mathbf{D_6}/\mathbf{Z_3}$. It is easy to calculate the quotients of the subgroups which contain $\mathbf{Z_3}$ because in all three cases, for the copy of $\mathbf{Z_6}$, and for the two copies of $\mathbf{D_3}$, the quotients are order 2 and so they must be $\mathbf{Z_2}$. The question remaining is how the three copies of $\mathbf{V}$ - which intersect $\mathbf{Z_3}$ only in the identity - are transformed as we take the quotient. We note that, for each copy of $\mathbf{V}$, $\mathbf{D_6} = \mathbf{V} \vee \mathbf{Z_3}$. Then $\mathbf{D_6}/\mathbf{Z_3} = \mathbf{V}/\mathbf{Z_3} \vee \mathbf{Z_3}/\mathbf{Z_3}$ by theorem 75a We have prior knowledge that $\mathbf{D_6}/\mathbf{Z_3} \simeq \mathbf{V}$ (it is order 4 and contains three copies of $\mathbf{Z_2}$) and so we have $\mathbf{V} = \mathbf{V}/\mathbf{Z_3} \vee \mathbf{1}$. Thus the copies of $\mathbf{V}$ that are contained in $\mathbf{D_6}$ are all merged into the final quotient group. The diagram given here is meant to illustrate the process, with distinct colors for each partial quotient. For clarity, the lattice diagram of $\mathbf{D_6}$ has been rearranged from the diagram previously given in lesson 25 of part 1.

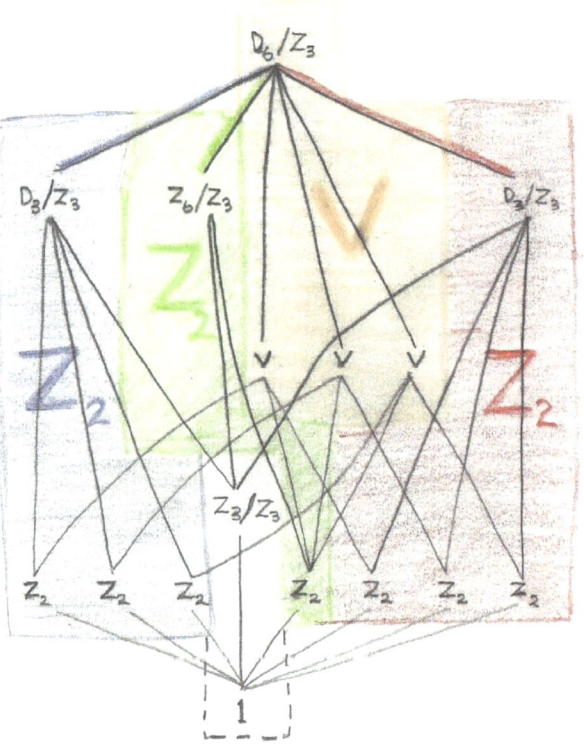

In general, if a subgroup meets the subgroup we are modding out, only their intersection will go into that part of the quotient. Naturally, in very large groups we cannot see the details of the quotient, but usually we will only need to know parts of the result.

Exercise 91: Carry out the above process to show the details of $\mathbf{D_8}/\mathbf{Z_2}$.

Exercise 92: Carry out the above process to show the details of $\mathbf{Z_{120}}/\mathbf{Z_6}$.

Exercise 93: Carry out the above process to show the details of $\mathbf{A_4}/\mathbf{V}$.

Exercise 94: Prove theorem 89 part c.

Exercise 95: Prove theorem 90 part b.

Lesson 24: The Second and Third Isomorphism Theorems

We have seen how the process of taking a quotient preserves the group structure, and how that is reflected in the lattice diagram of the group. The next few theorems go further. They uncover some unexpected isomorphisms between some of the many quotients of various subgroups. First there is a close relationship between a quotient involving the join of two subgroups and a quotient involving the intersection of those two subgroups. It is a general relationship requiring only that one of the subgroups is normal.

Theorem 91 (The Second Isomorphism Theorem): Let G be any group with $N \triangleleft G$ and $H \leq G$. Then we have $(HN)/N \simeq H/(H \cap N)$.

Proof: Since $N \triangleleft G$, we know $H \vee N = HN$ by theorem 61, and we also know that $H \cap N \triangleleft H$ by theorem 70. We will use this theorem as an opportunity to illustrate another technique for proving two seemingly distinct groups are isomorphic. Our standard method is to invent a candidate isomorphism and prove that it is one. This time we will define a candidate surjective homomorphism from HN to $H/(H \cap N)$, then show that the kernal of the homomorphism is N, and then appeal to the First Isomorphism Theorem to conclude that an isomorphism exists.

As the candidate homomorphism, we will define $\phi: HN \to H/(H \cap N)$ as simply as possible by $\phi(hn) = h(H \cap N)$. ϕ is clearly surjective. We must check that ϕ is well-defined; in this case, this means to check that the result of applying ϕ does not depend on the factorization of an element in HN. So suppose that there is an h' ∈ H and an n' ∈ N such that hn = h'n', but that $\phi(hn) \neq \phi(h'n')$; that is, that $h(H \cap N) \neq h'(H \cap N)$. Then $H \cap N \neq h^{-1}h'(H \cap N)$, which means that $h^{-1}h'$, clearly an element of H, can't equal any element of N. But we have assumed hn = h'n', which means $h^{-1}h' = nn'^{-1} \in N$, a contradiction. Hence ϕ is well-defined.

Now we need only show that ϕ preserves products:
$$\phi(h_1 n_1 \cdot h_2 n_2) = \phi(h_1 h_2 n_1' n_2) = h_1 h_2 (H \cap N) = h_1(H \cap N) h_2 (H \cap N) = \phi(h_1 n_1) \phi(h_2 n_2)$$
Therefore ϕ as defined above is a surjective homomorphism from HN onto $H/(H \cap N)$.

Now $\ker(\phi) = \{hn \in HN \mid \phi(hn) = \mathbf{1} \in H/(H \cap N)\}$. Thus hn ∈ $\ker(\phi)$ iff hn ∈ $H \cap N$. In particular h ∈ N. Thus $\ker(\phi) = N$.

Therefore, by the First Isomorphism Theorem, there exists an isomorphism from (HN)/N onto $H/(H \cap N)$ and the theorem is proved.

Corollary 91a: Let H and K be two subgroups of G, one of which is normal. Then [HK:K] and [HK:H] are relatively prime.

Proof: is left to the exercises (use theorem 54).

Thus in any sizable (non-simple) group there will be internal quotients involving joins and intersections of the subgroups, internal quotients that are isomorphic to each other. It is interesting that there is so much internal symmetry to the group structure; but not unexpected since groups are the primary tool mathematicians use to study symmetry. There is a kind of self-referential quality built into the nature of mathematics, and algebra reveals a small bit of it.

A lattice diagram illustrating The Second Isomorphism Theorem is given at the end of this lesson along with a diagram for the next theorem. We now turn to examine how taking consecutive quotients works.

Theorem 92 (The Third Isomorphism Theorem) Let $K \triangleleft H \triangleleft G$ and also assume $K \triangleleft G$. Then we have $(G/K) / (H/K) \simeq G/H$.

Proof: The most difficult part of this proof is keeping the notation clear. Here I will use boldface letters to indicate elements and subgroups resulting from modding out the subgroup K. In this way the conclusion of the theorem should be stated as **G/H** $\simeq$ G/H, which makes it look deceptively obvious. A common mnemonic device for remembering the theorem, if you need one, is that it looks like we are dividing by fractions. We will prove this using the same technique as for the previous theorem, by defining a candidate surjective homomorphism, finding its kernal, and applying the First Isomorphism Theorem.

So we define ϕ: **G** $\to$ G/H we let $\phi(\mathbf{g})$ = gH. To see ϕ is well-defined, choose a different representative of gK, say g'. Then g' = gk for some k $\in$ K, and $\phi(\mathbf{g'})$ = g'H = gkH. Since K is a subgroup of H, k $\in$ H. Thus g'H = gkH = gH and ϕ is well defined.

ϕ is clearly surjective.

Next check that ϕ preserves products:
$$\phi(\mathbf{x}\cdot\mathbf{y}) = xyH = xH\cdot yH = \phi(\mathbf{x})\cdot\phi(\mathbf{y})$$
Hence ϕ is a homomorphism.

The kernal of ϕ consists in all those cosets of **K** in **G** which map to the identity in G/H - that is, to H. Now $\phi(\mathbf{g}) \in$ H iff gH = H, iff g $\in$ H iff $\gamma_K(g) \in \gamma_K(H)$ iff **g** $\in$ **H**. Thus **ker**(ϕ) = **H**. Therefore, by the First Isomorphism Theorem, there exists an isomorphism from **G/H** onto G/H.

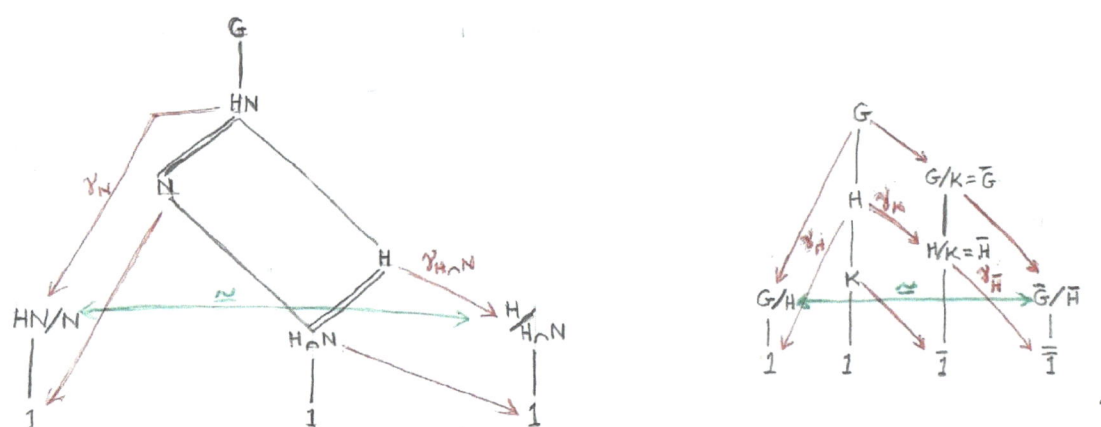

The Third Isomorphism Theorem can also be illustrated conveniently by a commutative diagram. It is an exercise for you to produce it.

Exercise 96: Prove Corollary 78a directly if possible. Otherwise use theorem 54.

Exercise 97: Construct a commutative diagram to illustrate the Third Isomorphism Theorem.

Lesson 25: The Zassenhaus Theorem

We now come to the last of our series of theorems on quotient groups. This is the most general of the theorems and is only fully useful in very large groups with a sufficient number of subgroups. You will find this in other algebra texts under the name Zassenhaus Lemma, but the term lemma is typically only applied to results that are not of interest in themselves but are only tools to get to something else that is of primary interest. I regard this result as interesting in itself, so I call it a theorem. It is also sometimes called the Butterfly Lemma because of the pattern in the lattice diagram that accompanies it. We will actually have occasion to apply this theorem, though not until book 3.

The full generality of this theorem requires that the group have two subgroups, each of which has a normal subgroup of its own. It requires that the group be large enough to accommodate these two pairs of subgroups and still have room for three other pairs of subgroups which come by taking joins and intersections of the various subgroups. From these rather minimalist initial conditions this theorem allows us to infer the existence of three internal quotients all mutually isomorphic to each other. We will state the theorem in three parts of a single theorem rather than dividing them into three theorems as I tend to do.

Theorem 93 (The Zassenhaus Theorem) Let H and K be any two subgroups of G and let $H^* \triangleleft H$ and $K^* \triangleleft K$. Then the following are true:

a) $H^*(H \cap K^*) \triangleleft H^*(H \cap K)$.
b) $K^*(H^* \cap K) \triangleleft K^*(H \cap K)$.
c) $(H^* \cap K)(H \cap K^*) \triangleleft H \cap K$.
d) $\dfrac{H^*(H \cap K)}{H^*(H \cap K^*)}$, $\dfrac{K^*(H \cap K)}{K^*(H^* \cap K)}$, and $\dfrac{H \cap K}{(H^* \cap K)(H \cap K^*)}$ are mutually isomorphic.

Proof a): Since $H^* \triangleleft H$, we know $H^* \vee (H \cap K) = H^*(H \cap K) \leq H$ by theorem 61. In the same way we know $H^* \vee (H \cap K^*) = H^*(H \cap K^*) \leq H$. Using theorem 61 again, since $K^* \triangleleft K$, we know that $K^* \vee (H \cap K) = K^*(H \cap K) \leq K$ and that $K^* \vee (H^* \cap K) = K^*(H^* \cap K) \leq K$.

In theorem 70 if we take G as H, H as $H \cap K$, and K as H^*, since $H^* \cap H \cap K = H^* \cap K$, we can conclude $H^* \cap K \triangleleft H \cap K$. Similarly in theorem 70 if we take G as K, H as $H \cap K$, and K as K^*, we can conclude that $H \cap K^* \triangleleft H \cap K$.

Finally in theorem 69 take H as $H \cap K^*$, S as $H \cap K$, K as H^*, and G as H; then the facts $H^* \triangleleft H$ and $H \cap K^* \triangleleft H \cap K$ allow us to conclude that $H^* \vee (H \cap K^*) \triangleleft H^* \vee (H \cap K)$. This is the same as $H^*(H \cap K^*) \triangleleft H^*(H \cap K)$ which is what we intended to show.

b): This is similar to part a and is left as an exercise.

c): In the proof of part a we showed that $H^* \cap K \triangleleft H \cap K$ and $H \cap K^* \triangleleft H \cap K$. Therefore by theorem 68b we conclude $(H^* \cap K)(H \cap K^*) \triangleleft H \cap K$, where we take $H \cap K$ in place of G in the statement of that theorem.

d): We will prove that the first of the quotients is isomorphic to the third of the quotients. Let's begin by simplifying the notation: let $L = (H^* \cap K)(H \cap K^*)$. First we define a candidate homomorphism

$$\phi: H^*(H \cap K) \to (H \cap K)/L$$

by this rule: $\phi(h^*x) = xL$ where $h^* \in H^*$ and $x \in H \cap K$.

To see that ϕ is well-defined, note that it depends only on the choice of x. So let's choose distinct elements $x_1, x_2 \in H \cap K$ and distinct elements $h_1^*, h_2^* \in H^*$ such that $h_1^* x_1 = h_2^* x_2$. This means $x_1 x_2^{-1} = h_1^{*-1} \cdot h_2^*$. Therefore $x_1 x_2^{-1} \in H^*$; so $x_1 x_2^{-1} \in H^* \cap (H \cap K) = H^* \cap K \leq L$. Thus x_1 and x_2 are in the same coset of L in $H \cap K$, and hence $\phi(h_1^* x_1) = \phi(h_2^* x_2)$. Therefore ϕ is well-defined.

φ is clearly surjective.

Since $H^* \triangleleft H$ and $x_1 \in H \cap K$ and thus $x_1 \in H$, there is some $h_3^* \in H^*$ such that $x_1 h_2^* = h_3^* x_1$. Using this we compute:

$$\phi[(h_1^* x_1)(h_2^* x_2)] = \phi[h_1^*(x_1 h_2^*) x_2] = \phi[h_1^*(h_3^* x_1) x_2] = \phi[(h_1^* h_3^*)(x_1 x_2)] = x_1 x_2 L$$
$$= x_1 L x_2 L = \phi(h_1^* x_1) \phi(h_2^* x_2)$$

Thus φ preserves products and is a homomorphism.

To compute **ker**(φ), note that $h^* x \in $ **ker**(φ) iff $\phi(h^* x) = xL = L$. Hence x must be in L and therefore $h^* x$ must be in $H^* L = H^*(H^* \cap K)(H \cap K^*) = H^*(H \cap K^*)$. Hence, by the First Isomorphism Theorem, there exists an isomorphism from $[H^*(H \cap K)] / [H^*(H \cap K^*)]$ onto $(H \cap K)/L$, which is what we were to prove.

The rest of the proof is similar and is left as an exercise.

We now have enough tools for handling quotients internal to the group, though the exact applications of some of these theorems may be tricky to recognize, as they were in the proof of the Zassenhaus Theorem. It is frequently helpful to draw a partial lattice diagram to recognize when some of our previous theorems are applicable. Using these we can accomplish a great deal when dealing with very large groups. There is one more related idea that concerns internal relationships between subgroups that is the subject of the next lesson.

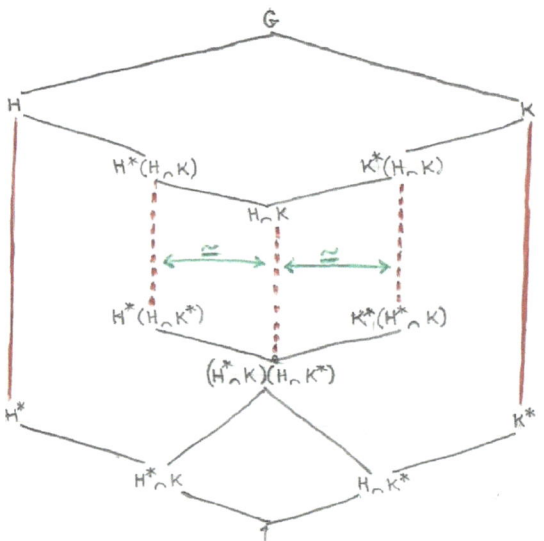

Exercise 98: Prove theorem 93 part b.

Exercise 99: Prove that the second quotient is isomorphic to the third quotient in theorem 93 part d, and thus finish the theorem.

Exercise 100: Restate the Zassenhaus Theorem for the following special cases and give the corresponding lattice diagram:

a) $H \cap K = 1$
b) $H^* \cap K = 1$ but $H \cap K^* \neq 1$
c) both $H^* \cap K$ and $H \cap K^* = 1$
d) $H^*, K^* \leq H \cap K$
e) $H^* = K^*$
f) $H \leq K$ but $H^* \neq K^*$

Hans Zassenhaus was born in 1912 in Koblenz, Germany. His father was an admirer of Albert Schweitzer and lost his job because of it when the Nazi's came to power. Zassenhaus went to the University of Hamburg intending to become a physicist but was converted to mathematics when he attended a course taught by Emil Artin. While still a student he proved the Butterfly Lemma and published a book, *The Theory of Groups*, based on Artin's lectures. His thesis in 1934 was on doubly transitive groups. After the war, in 1949 he moved to McGill University in Montreal, to Notre Dame in 1959, and finally to Ohio State University in 1963 where he remained until he retired. While visiting CalTech he helped pioneer the use of computers in the study of number theory. He died in 1991 in Columbus, Ohio.

Lesson 26: Normalizers

The last four lessons have focused on relative normality; I mean, the normality of one subgroup within another subgroup that contains it and not necessarily normality in the whole group. When we have such relative normality within a subgroup, then we have deduced several relative quotients between those subgroups which turned out to be isomorphic to each other. Normality is one way we obtain a sort of pseudo-commutativity, so when a subgroup is not normal we are naturally interested in its normality relative to another subgroup. We want to get as much normality as possible. This leads naturally to the following idea, the largest subgroup in which our particular subgroup is normal.

Definition 57: Let $H \leq G$ be an arbitrary subgroup. The set $\mathbf{N}(H) = \{g \in \mathbf{G} \mid g^{-1}Hg = H\}$ is called the **normalizer** of H in G.

The following properties of the normalizer are easy to prove and are left as exercises.

Theorem 94: a) $\mathbf{N}(H) \leq G$.
       b) $\mathbf{N}(H)$ is the largest subgroup of G in which H is normal; or, in other words, if $H \triangleleft K$ then $K \leq \mathbf{N}(H)$.
       c) if G is Abelian and H is any subgroup of G, then $\mathbf{N}(H) = G$.
       d) if $H \triangleleft G$, then $\mathbf{N}(H) = G$.

Proof: is left to the exercises.

The normalizer of H will not contain any subgroups of G which are conjugate to H. Since $H \triangleleft \mathbf{N}(H)$, H has no conjugate subgroup within $\mathbf{N}(H)$. All of the conjugates of H in G must arise from conjugation by elements that are not in $\mathbf{N}(H)$. Note that $\mathbf{N}(H)$ is never trivial because H is always normal in itself. Hence the other extreme from a normal subgroup is the following:

Definition 58: Let $H \leq G$ be an arbitrary subgroup. When $\mathbf{N}(H) = H$ we say that H is a **self-normalizing** subgroup of G.

It is in regard to the conjugates of a particular subgroup that the normalizer is most helpful. We always benefit when we can count how many of a particular kind of object exists. If H is normal in G, then it has no conjugate subgroups. But if H is not normal in G, how many conjugate subgroups does it have?

Theorem 95: The number of conjugate subgroups to H in G equals $[G:\mathbf{N}(H)]$.

Proof: Instead of asking when two conjugates of H are distinct, let's come from the opposite direction and ask when two conjugates of H are the same. $x^{-1}Hx = y^{-1}Hy$ iff $H = xy^{-1}Hyx^{-1}$ which is true iff $H = (yx^{-1})^{-1}Hyx^{-1}$. In other words $x^{-1}Hx = y^{-1}Hy$ iff $yx^{-1} \in \mathbf{N}(H)$. Thus $x^{-1}Hx = y^{-1}Hy$ iff $y \in \mathbf{N}(H)x$; that is, x and y are in the same right coset of $\mathbf{N}(H)$. Therefore two conjugates of H are distinct iff the conjugating elements come from distinct cosets of $\mathbf{N}(H)$.

As an example of this, consider the subgroup $H = \langle \delta_1 \rangle$ in $\mathbf{D_6}$. Thus H is one of the copies of $\mathbf{Z_2}$, one of the copies that is not normal in $\mathbf{D_6}$. Referring back to the lattice diagram for this group it should be clear that the largest subgroup of $\mathbf{D_6}$ in which H is normal is the copy of $\mathbf{V}$ containing it. Since $[\mathbf{D_6}:\mathbf{V}] = 3$, theorem 95 tells us that there are a total of three subgroups conjugate to $\langle \delta_1 \rangle$ (this includes $\langle \delta_1 \rangle$ itself as one of its conjugates.) The copy of $\mathbf{V}$ in question includes ρ_0, ρ_3, δ_1, and μ_3. Conjugation of δ_1 by any element in this copy of V yields δ_1 or ρ_0, the two elements of $\langle \delta_1 \rangle$.

Will distinct subgroups which are conjugate to each other have distinct normalizers? $\mathbf{N}(H)$ could conceivably normalize a subgroup conjugate to H as well. The subgroup conjugate to H is conjugate to it by some element outside of $\mathbf{N}(H)$. It would be handy to know how many of the conjugates of H are normalized by $\mathbf{N}(H)$ and how to find the normalizers of the ones that aren't. The next theorem tells all we need to know.

Theorem 96: $N(g^{-1}Hg) = g^{-1}N(H)g$.
Proof: We will show that each side of the equation is a subgroup of the other.
  Let $n \in N(g^{-1}Hg)$. Then by definition $n^{-1}g^{-1}Hgn = g^{-1}Hg$. Then solving the right hand side for H, we have $gn^{-1}g^{-1}Hgng^{-1} = H$. Hence $gng^{-1} \in N(H)$. Therefore $n \in g^{-1}N(H)g$. Thus we have shown that $N(g^{-1}Hg) \leq g^{-1}N(H)g$.
  Now suppose $n \in N(H)$ and consider $g^{-1}ng$. We will show that $g^{-1}ng$ normalizes $g^{-1}Hg$:
$$(g^{-1}ng)^{-1} \cdot g^{-1}Hg \cdot (g^{-1}ng) = g^{-1}n^{-1}g \cdot g^{-1}Hg \cdot g^{-1}ng = g^{-1}n^{-1}Hng = g^{-1}Hg$$
where the last step is justified because we assumed n normalized H. Thus we have shown that $g^{-1}N(H)g \leq N(g^{-1}Hg)$, and so the two sides are equal.

Corollary 96a: Let $H \leq G$. If $N(H) \triangleleft G$, then $N(H)$ contains all the subgroups conjugate to H.
Proof: is an exercise.

Let's continue discussing D_6. Each conjugate of the normalizer of $<\delta_1>$, $V = \{\rho_0, \rho_3, \delta_1, \mu_3\}$, determines one of the two conjugates of $<\delta_1>$. For example, take the conjugate of V by δ_2:
$$\delta_2 V \delta_2 = \{\rho_0, \rho_3, \delta_3, \mu_2\} = N(\delta_2 <\delta_1> \delta_2) = N(<\delta_3>).$$
So conjugation of δ_3 by any element of $\{\rho_0, \rho_3, \delta_3, \mu_2\}$ will yield either δ_3 or ρ_0. Thus theorem 95 gives us a way to find the conjugate subgroups of H, as well as the normalizers of each conjugate of H, as well as count them. The elements δ_1, δ_2, and δ_3 are a complete set of mutually conjugate elements, each one normalized by one of the copies of V. The elements μ_1, μ_2, and μ_3 are also a complete set of mutually conjugate elements, also normalized by the copies of V. None of the μ's are conjugate to any of the δ's. What of the other element of order 2, namely ρ_3? We know that $<\rho_3> \triangleleft D_6$. Hence the number of conjugates of ρ_3 is exactly one. This gives rise to new terminology:

Definition 59: Let $g \in G$ be any element of the group G. The set $\{x^{-1}gx \mid x \in G\}$ is called the **conjugacy class** of g and is denoted by $C(g)$.

Thus D_6 has three conjugacy classes of elements of order 2, of sizes 3, 3 and 1. Recall that by theorem 72, in S_n the number of permutations conjugate to a particular permutation σ are all the permutations of the same cycle type as σ. This will be useful to remember in exercise 102.

  We will also define a relative normalizer. If $H \leq K \leq G$, then it makes sense to talk about the normalizer of H within K and without regard to G. It is simply the subgroup of all the elements of K which normalize H. We will denote such normalizers relative to another subgroup by a subscript. Thus the normalizer of H in K is denoted by $N_K(H)$. Unless there is danger of confusion, we will not use a subscript for the normalizer within the whole group G.

Theorem 97: Let $H \leq K \leq G$. Then $N_K(H) = K \cap N(H)$.
Proof: is an exercise.

Exercise 101: Prove theorem 94.
Exercise 102: Prove corollary 96a.
Exercise 103: Prove theorem 97.
Exercise 104: Refer back to exercise 14 in lesson 4. Find all the normalizers of elements of order 2, and the conjugacy classes of the elements of order 2.
Exercise 105: In S_6 let σ have the cycle type 1,1,1,3. How many permutations are conjugate to σ? How many subgroups conjugate to $<\sigma>$ are there? What is the index of $N(<\sigma>)$? Can you determine which subgroups are normalizers of permutations of this cycle type?
Exercise 106: If H and K are two subgroups of G and $H \leq N(K)$, then $HK \leq G$.

Lesson 27: The Center of a Group

In the remaining lessons of part 2 we will discuss three particularly important subgroups that are defined in every group and whose properties are a great aid in analyzing groups in general.

Definition 60: Let G be any group. The set $\{z \in G \mid zg = gz \ \forall g \in G\}$ is called the **center** of G and is denoted by **Z**(G).

The center is the set of all elements of G that commute with every other element of G. We know that the identity element is one of these, so we always have $\mathbf{1} \in \mathbf{Z}(G)$. The elements can be easily recognized if we have a Cayley table for the group: an element is in the center of G if the column it determines is identical to the row it determines. As you might expect, the center of G is indeed a subgroup of G. In fact, it is a special subgroup.

Theorem 98: Let G be any group. Then $\mathbf{Z}(G) \ll G$.

Proof: The proof that $\mathbf{Z}(G) \leq G$ is routine and is left as an exercise.

To show that $\mathbf{Z}(G)$ is a characteristic subgroup, choose an arbitrary element $\beta \in \mathbf{Aut}(G)$, and an arbitrary element $z \in \mathbf{Z}(G)$. Then by definition $zg = gz$ for all $g \in G$. Applying β to both sides we have $\beta(zg) = \beta(gz)$. Then $\beta(z)\beta(g) = \beta(g)\beta(z)$ for all g. As g varies over the whole of G, $\beta(g)$ varies over all of G because β is a bijection. Thus $\beta(z)$ commutes with every element of G and $\beta(z) \in \mathbf{Z}(G)$ by definition. Since β preserves $\mathbf{Z}(G)$ and was an arbitrary automorphism, we have shown that $\mathbf{Z}(G)$ is a characteristic subgroup of G.

Now recall that elementary groups have no *proper* characteristic subgroups. This means that for an elementary Abelian group the center equals the whole group, which it would in any case for an Abelian group of course. In the case of an elementary non-Abelian group, this means that the center must be trivial. We will make use of the following obvious, but not strictly accurate, terminology:

Definition 61: The group G whose center is trivial is called a **centerless group**.

Centerless groups are necessarily non-Abelian; by this measure, one could say that centerless groups are the least Abelian of all groups. This is not the first time we have remarked that a certain class of groups was the "least Abelian". It is not a precise idea.

Corollary 98a: If $H \triangleleft G$, then $\mathbf{Z}(H) \triangleleft G$.

Proof: hint, use theorem 66.

Since the center is a characteristic subgroup it is also a normal subgroup and we can form the quotient. Forming this quotient we uncover an unexpected connection.

Theorem 99: For any group G, $G/\mathbf{Z}(G) \simeq \mathbf{Inn}(G)$.

Proof: We will abbreviate $\mathbf{Z}(G)$ to simply Z. First, as usual, we will define a candidate homomorphism $\phi : G \to \mathbf{Inn}(G)$. However, if we do this in the most straight-forward manner, $\phi(g) = \kappa_g$, we find that we get an anti-homomorphism, a homomorphism which reverses the order of multiplication:

$$\phi(gh) = \kappa_{gh}(x) = (gh)^{-1}x(gh) = h^{-1}g^{-1}xgh = h^{-1}(g^{-1}xg)h = \kappa_h(\kappa_g(x)) = \phi(h)\phi(g)$$

Anti-homomorphisms are as valid as regular homomorphisms, but they are irritating aesthetically. To avoid them, we will use a slightly more complicated definition for ϕ, namely $\phi(g) = \kappa_{g^{(-1)}}$; then we have:

$$\phi(gh) = \kappa_{(gh)^{(-1)}}(x) = (gh)x(gh)^{-1} = ghxh^{-1}g^{-1} = g(hxh^{-1})g^{-1} = \kappa_g(\kappa_h(x)) = \phi(g)\phi(h)$$

ϕ is clearly well-defined and surjective and it preserves products.

Thus ϕ is a homomorphism. $\mathbf{ker}(\phi)$ is the set of all $g \in G$ for which $\kappa_{g^{(-1)}} = \mathbf{1}_G$, the identity automorphism. Now that $\kappa_{g^{(-1)}}(x) = x$ for all x means that $gxg^{-1} = x$ for all x. Thus $gx = xg$ for all x. By definition then $g \in \mathbf{Z}(G)$ and $\mathbf{ker}(\phi) = \mathbf{Z}(G)$. Therefore, by the First Isomorphism Theorem, there exists an isomorphism from $G/\mathbf{Z}(G)$ onto $\mathbf{Inn}(G)$.

This gives us a relatively easy way to calculate **Inn**(G). It is generally easier to find the center of a group than to calculate all the conjugation maps one at a time. It will be helpful to know the inner automorphism groups of some of the families that we know. Of course we do know that **Inn**(G) = 1 for any Abelian group, so it is only non-Abelian groups that we need any help with. The following are easy to see when $n \geq 4$:

$$\mathbf{Z}(\mathbf{S_n}) = 1 \implies \mathbf{Inn}(\mathbf{S_n}) \simeq \mathbf{S_n}$$
$$\mathbf{Z}(\mathbf{A_n}) = 1 \implies \mathbf{Inn}(\mathbf{A_n}) \simeq \mathbf{A_n}$$

We can use the presentation to compute $\mathbf{Z}(\mathbf{D_n})$. All we need do is examine the commutativity properties of the generators. $\mathbf{D_n} = <a, r : a^2 = 1, r^n = 1, ar = r^{-1}a>$. The last relation implies $ar^k = r^{-k}a$. The generator, a, will commute with a power of r only if r^k is its own inverse. That happens only when $k = n/2$. $r^{n/2}$ is the element of $\mathbf{D_n}$ usually interpreted as the 180^0 rotation, and it only exists if n is even. Thus we have:

$$\mathbf{Z}(\mathbf{D_n}) = 1 \text{ if n is odd} \implies \mathbf{Inn}(\mathbf{D_n}) \simeq \mathbf{D_n} \text{ if n is odd}$$
$$\mathbf{Z}(\mathbf{D_n}) = \mathbf{Z_2} \text{ if n is even} \implies \mathbf{Inn}(\mathbf{D_n}) \simeq \mathbf{D_n}/\mathbf{Z_2} \text{ if n is even}$$

One other family of groups we know by their presentation are the generalized quaternion groups. $\mathbf{Q}_{2^{\wedge}(n+1)} = <x, y : x^{2^{\wedge}n} = 1, y^4 = 1, x^{2^{\wedge}(n-1)} = y^2, xy = yx^{-1}>$. Again, the power of x which commutes with all powers of y is the power of x that is its own inverse. This is exactly $x^{2^{\wedge}(n-1)} = y^2$. Now the subgroup $<y^2> = \mathbf{Z_2}$. Thus

$$\mathbf{Z}(\mathbf{Q}_{2^{\wedge}(n+1)}) = \mathbf{Z_2} \implies \mathbf{Inn}(\mathbf{Q}_{2^{\wedge}(n+1)}) \simeq \mathbf{Q}_{2^{\wedge}(n+1)}/\mathbf{Z_2}$$

In lesson 9 we found all of the automorphisms, inner and outer, with a total of 24 of them including the identity automorphism. From the above calculation, we expect a total of 4 inner automorphism, including the identity. The outer automorphisms are distinct up to a coset of the inner automorphism group, so we expect a total of 5 outer automorphisms beside the representative of the inner automorphism subgroup. Use your work on exercise 78 of lesson 20 to check your answer.

Theorem 100: Let H be any subgroup of G. Then $\mathbf{Z}(G) \leq \mathbf{N}(H)$.

Proof: Let $z \in \mathbf{Z}(G)$, and let $h \in H$ be arbitrary. Then $zh = hz$ by definition of the center. This implies $h = z^{-1}hz$. Therefore $z \in \mathbf{N}(H)$ and we have shown that $\mathbf{Z}(G) \leq \mathbf{N}(H)$.

Since H was an arbitrary subgroup of G we see that the center of a group is contained in each and every normalizer in G. Thus we immediately see that

Corollary 100a: $\mathbf{Z}(G) \leq \bigcap_{H \leq G} \mathbf{N}(H)$.

Theorem 101: If $G/\mathbf{Z}(G)$ is cyclic, then $\mathbf{Z}(G) = G$.

Proof: Suppose that $G/\mathbf{Z}(G)$ is cyclic. Then there is some coset of $\mathbf{Z}(G)$ (which we abbreviate as Z) that generates every coset in G/Z. Let gZ be that generator. Let x be an arbitrary element of G and consider the element xg. xg is in some coset of Z, so it is in some coset of the form g^kZ. Hence we can write $xg = g^k z$ for some $z \in Z$. Then $x = g^k z g^{-1} = zg^k g^{-1} = zg^{k-1}$. Thus we have $gx = gzg^{k-1} = g^k z$. Hence $xg = gx$. Thus g commutes with an arbitrary element of G and therefore by definition $g \in Z$. We have shown if G/Z is cyclic then $G = Z$.

In other words, $G/\mathbf{Z}(G)$ can only be cyclic when G itself is cyclic and then $\mathbf{Z}(G) = G$. But the converse is not true: $\mathbf{Z}(G) = G$ whenever G is Abelian. Immediately we get the corollary:

Corollary 101a: If G is non-Abelian, **Inn**(G) is never a non-trivial cyclic group.

There are two more results to complete this lesson, both exercises.

Theorem 102: $\mathbf{Z}(G \times H) \simeq \mathbf{Z}(G) \times \mathbf{Z}(H)$.

Proof: is an exercise.

Theorem 103: If $[G:Z(G)] = n$, then for any $g \in G$, $|C(g)| \leq n$.
Proof: is an exercise.

    Thus the center of a group is another measure of how Abelian a group is. When the center equals the entire group, then the group is Abelian. By this measure, the next nearest groups to being Abelian are those whose centers have index 2; half of the elements of those groups commute with every other element. The symmetric groups and alternating groups on more than 3 letters, and the dihedral groups with an odd number of vertices are centerless and as such are among the least Abelian groups. The dihedral groups with an even number of vertices and the generalized quaternions both are families of groups whose centers are very nearly trivial. It is natural that in studying non-Abelian groups we would be led to consider the center of the group. We are always trying to approximate a non-Abelian group as nearly as possible by an Abelian piece of it. Soon we will look at one more approach to doing this same thing which gives yet a different way of measuring Abelian-ness.

Exercise 107: Complete the proof of theorem 98 by showing that $Z(G) \leq G$.
Exercise 108: Prove corollaries 98a, 100a and 101a.
Exercise 109: Prove theorem 102.
Exercise 110: Prove theorem 103.

Lesson 28: The Centralizer and the Conjugacy Class Equation

We defined the normalizer for subgroups only. It is sometimes generalized to apply to subsets and not to subgroups alone, but I will not be using it in that generality. Now, however, when we proceed in a similar fashion from centers to centralizers, I will generalize it to apply to subsets and individual elements, and will rarely use it with regard to subgroups specifically. Centralizers have the same general use and value with respect to elements that normalizers have with respect to subgroups.

Definition 62: Let $S \subseteq G$ be any subset of the group G. The set $\{x \in G \mid xg = gx \ \forall g \in S\}$ is called the **centralizer** of S and is denoted by $Z_G(S)$.

We begin by proving the same general properties for the centralizer that we proved for the normalizer. If $S = \{g\}$ we will write $Z_G(g)$ rather than $Z_G(\{g\})$. The subscript is only used with the centralizer and not with the center.

Theorem 104: Let $S \subseteq G$ be any subset of the group G. Then the following are true:
  a) $Z_G(S) \leq G$
  b) $<g> \triangleleft Z_G(g) \leq N(<g>)$
  c) $Z_G(g, h) = Z_G(g) \cap Z_G(h)$
  d) If $S \subseteq T$ then $Z_G(T) \leq Z_G(S)$

Proof: the proofs of each part are routine and are left as exercises.

An important use of centralizers for us will be in counting the number of conjugate elements that a particular element has. Just as a particular subgroup has as many conjugate subgroups as the index of the normalizer, so a particular element has as many conjugate elements as the index of the centralizer. For the element g, the number of elements conjugate to g is the order of the conjugacy class of g, $|C(g)|$. It is clear that if $g \in C(h)$ then $h \in C(g)$. The conjugacy classes are not themselves subgroups of G, but they do form a partition of the elements of G.

Theorem 105: $|C(g)| = [G:Z_G(g)]$

Proof: Abbreviate $Z_G(g)$ as Z. Let $x \in Zy$, so that $x = zy$ for some $z \in Z$. Then we can compute
$$x^{-1}gx = (zy)^{-1}g(zy) = y^{-1}z^{-1}gzy = y^{-1}z^{-1}zgy = y^{-1}gy$$
Hence elements from the same right coset of Z determine the same conjugate of g. Hence
$$|C(g)| \geq [G:Z_G(g)].$$
Now suppose $x^{-1}gx = y^{-1}gy$. Then $gxy^{-1} = xy^{-1}g$. Then $xy^{-1} \in Z$ and hence $x \in Zy$. Thus if two elements generate the same conjugation map, they are in the same coset of Z. Hence we have $|C(g)| \leq [G:Z_G(g)]$. Therefore $|C(g)| = [G:Z_G(g)]$.

Now if $g \in Z(G)$, then g has no conjugates other than itself and $|C(g)| = 1$. Since the conjugacy classes partition G, if we add up the sizes of all the conjugacy classes then the total is the order of G. This gives us the following:

Theorem 106: (The Conjugacy Class Equation) Let $g_1, g_2, \ldots, g_n$ be a complete set of representatives of the conjugacy classes of non-central elements of G. Then we have
$$|G| = |Z(G)| + \sum_{i=1}^{n} [G:Z_G(g_i)].$$

Proof: is an easy exercise.

It is typical that we always get more than we expect when we learn how to count something. We will now look at one application of the conjugacy class equation in this lesson and consider others in the next lesson. Recall that Lagrange's Theorem stated that the order of a subgroup always divides the order of the group, but that in general the converse is not true. There are groups, A_4 being an example, for which there are divisors of the order of the group with no corresponding subgroup of that order; A_4 is order 12 but has no subgroup of order 6. We can now get a *partial* converse to Lagrange's Theorem.

Theorem 107 (Cauchy's Theorem): Suppose the prime p divides the order of the group G. Then G has an element of order p.

Proof: If G has no proper subgroups at all, then it is prime cyclic and we are done. Hence we can let $|G| = mp$ for $m > 1$. We know the theorem holds for $m = 1$ and we will use induction on m. We will assume the theorem is true for all G such that $|G| = np$ and $n < m$. There are two cases.

Case 1: Suppose G contains a proper subgroup H whose index $[G:H]$ is not divisible by p. Then, $|H| = kp$ and $k < m$. By induction H contains an element of order p, and hence G contains an element of order p.

Case 2: Now suppose every proper subgroup of G has an index divisible by p. Write the conjugacy class equation:

$$|G| = |\mathbf{Z}(G)| + \sum_{i=1}^{n} [G:\mathbf{Z}_G(g_i)] = n_0 + n_1 + \ldots + n_k$$

where $n_0 = |\mathbf{Z}(G)|$ and each of the n_i's after n_0 equals an index of a subgroup and is therefore divisible by p. Hence n_0 must also be divisible by p. By induction, $\mathbf{Z}(G)$ includes an element of order p and hence G also includes an element of order p.

Exercise 111: Prove each part of Theorem 104.

Exercise 112: Give a formal proof of the Conjugacy Class Equation.

Exercise 113: Let H be any subgroup of G and $|H| = 2$. Then $\mathbf{Z}_G(H) = \mathbf{N}(H)$.

Exercise 114: Show that for any element $s \in G$, $g^{-1}\mathbf{Z}_G(s)g = \mathbf{Z}_G(g^{-1}sg)$.

Exercise 115: Let $H \triangleleft G$, then for $g \in G$ $\kappa_g|_H \in \mathbf{Aut}(H)$. Hence there is an induced homomorphism $\phi: G \to \mathbf{Aut}(H)$ such that $\ker(\phi) = \mathbf{Z}_G(H)$. Define ϕ explicitly and conclude that $G/\mathbf{Z}_G(H)$ is isomorphic to a subgroup of $\mathbf{Aut}(H)$.

Exercise 116: Continuing exercise 114, argue that $\mathbf{N}(H)/\mathbf{Z}_G(H)$ is isomorphic to a subgroup of $\mathbf{Aut}(H)$.

Exercise 117: If $\beta \in \mathbf{Aut}(G)$ – that is, β is a permutation of the elements of G – show that β permutes the conjugacy classes of G.

Augustin Louis Cauchy was born in 1789 to a staunchly royalist and Catholic family. They had to flee Paris to escape the reign of terror but returned to Paris once Bonaparte was in charge. In Paris, Cauchy's father worked for Laplace, a famous mathematician, and Lagrange was a family friend. With their encouragement he pursued the study of engineering but after four years actual work as an engineer he became bored with it and turned to mathematics. In 1816, after Napoleon was deposed, Cauchy was promoted by the new king, over the opposition of many of his colleagues. In 1818 his father arranged a marriage for him to Aloise de Bure, and it was as happy as many marriages. In 1830 he and his family fled Paris again when Louis-Philippe became king and spent eight years in various European universities. In 1838 the political climate changed again and he returned to Paris but was unable to secure a university position due to his refusal to take an oath of loyalty to the government. Instead the Catholic church hired him to help establish a system of Catholic schools and universities in France. In 1848 the oath of allegiance was abolished and he could then teach at secular universities. His political views always caused conflict with his colleagues. Abel, in particular, had harsh words to say about him. He is second only to Euler in the number of mathematical papers he wrote, and has more theorems and formulas named for him than any other mathematician. His most important achievement is perhaps as the founder of complex function theory. He died in 1857 and is one of only 72 people whose names are inscribed on the Eiffel Tower.

Lesson 29: p-Groups

We will now consider a class of groups which will prove central to our lessons in part 3. Here we will consider their basic properties.

Definition 63: Let G be a group of order p^n for a prime p and $n \geq 1$. Such a group will be called a **p-group**.

Note that every subgroup of a p-group, and every quotient of a p-group, is also a p-group. This is an easy thing to prove and is left as an exercise. We know that $|Q_{2^{\wedge}(n+1)}| = 2^{n+1}$ so the generalized quaternions are all 2-groups. D_4, D_8, D_{16}, etc. are all 2-groups. Prime cyclic groups, of course, are all p-groups and all elementary Abelian groups are also p-groups. Any group of order 121 is an 11-group. The p-subgroups of a given group turn out to be nearly as important for understanding the structure of the group as the cyclic subgroups. The Conjugacy Class Equation is the most important tool for analysing p-groups.

Theorem 108: Let P be a p-group. Then $Z(P) \neq 1$.
Proof: For G the conjugacy class equation is

$$|G| = |Z(G)| + \sum_{i=1}^{n} [G:Z_G(g_i)]$$

where the g_i are representatives of the distinct conjugacy classes of non-central elements. Since the g_i are non-central, $Z_G(g_i) \neq G$ for any of the g_i. Then we know that p divides each of the indices $[G:Z_G(g_i)]$. Since p divides $|G|$ it follows that p must also divide $|Z(G)|$ and hence $Z(G) \neq 1$.

Theorem 109: Let P be a p-group, and let H be a non-trivial normal subgroup of P. Then $H \cap Z(P) \neq 1$.
Proof: Let H be a non-trivial normal subgroup of P, and H is clearly a p-group in its own right. That H is normal means that H includes all the conjugacy classes of its own elements. So if C is a conjugacy class of an element of P either $C \subseteq H$ or else $C \cap H = 1$. Choose representatives of each conjugacy class in P, $\{g_1, g_2, ..., g_n\}$ including the elements that are in $Z(P)$, whose conjugacy classes have size 1. Renumbering if necessary, let $C(g_1), C(g_2), ..., C(g_k)$ be the conjugacy classes contained in H. Of these let $g_1, g_2, ..., g_j \in Z(P)$ and hence in $H \cap Z(P)$. Then we can write the Conjugacy Class Equation for H:

$$|H| = |H \cap Z(P)| + \sum_{i=1}^{n} [P:Z_P(g_i)]$$

We know that p divides the left hand side and that p divides each term in the summation, so p must divide $|H \cap Z(P)|$. Therefore $H \cap Z(P) \neq 1$.

Corollary 109a: Let P be a p-group and $H \triangleleft P$ with $|H| = p$. Then $H \leq Z(P)$.
Proof: is immediate.

The center contains all the normal subgroups of order p, In fact, it is not too hard to argue the following:

Corollary 109b: $soc(P) \leq Z(P)$.
Proof: is an exercise.

The center contains all the minimal normal subgroups, and in particular all the normal subgroups of order p. However, for all we know at the moment there may be minimal normal subgroups of larger order. The next theorem shows that there are no minimal subgroups that are larger than order p. Take care to realize that there may be plenty of copies of Z_p which are not normal in the whole group. D_4, for example, has four copies of Z_2 that are not normal in D_4.

Theorem 110: Let P be a p-group and let H be any normal subgroup of P. Then H contains a subgroup of order p^k for every k such that p^k divides the order of H, and this subgroup is normal in P.

Proof: Let $|H| = p^m$ and H normal in P. We will prove this by induction on m. If m = 1 or 0 the result is trivial. So take m > 1 and suppose the theorem is true for all n < m. By theorem 109, $H \cap \mathbf{Z}(P) \neq \mathbf{1}$, and by Cauchy's Theorem $H \cap \mathbf{Z}(P)$ contains a subgroup K of order p. Since $K \leq \mathbf{Z}(P)$, K is also normal in P. Then the quotient group P/K has order p^{m-1} and by the First Correspondence Theorem H/K ◁ P/K. By induction, H/K has a subgroup of every order $p, p^2, \ldots, p^{m-2}$ and these are all normal in P/K. Mapping back to P these correspond to subgroups of H of orders $p^2, \ldots, p^{m-1}$ each of which is normal in P.

Corollary 110a: If P is a p-group, then P has a normal subgroup of every order dividing the order of P.
Proof: Let H = P in theorem 110.

Thus a p-group P has a tower of normal subgroups $\mathbf{1} \triangleleft P_1 \triangleleft P_2 \triangleleft \ldots \triangleleft P_n = P$ where each P_i has order p^i and each $P_i \triangleleft P$. Such towers of subgroups, which we will call *series* of subgroups rather than towers, will play a critical role in part 3. There is one more theorem to round out our knowledge of the role of normality in p-groups.

Theorem 111: Let P be a p-group. If H is a proper subgroup of P, then H is a proper subgroup of its normalizer $\mathbf{N}(H)$.

Proof: We will again use induction, this time on the order of P. The result is trivial if P is Abelian, and therefore we can assume $|P| > p$. So we will assume the theorem is true for all P whose orders are less than p^m. Let H be a proper subgroup of P. By theorem 100 we know $\mathbf{Z}(P) \leq \mathbf{N}(H)$. We know $\mathbf{Z}(P)$ is non-trivial. If $\mathbf{Z}(P)$ is not contained in H, then H is properly contained in $\mathbf{Z}(P) \vee H$, which is contained in $\mathbf{N}(H)$ and the theorem holds. Hence we may assume $\mathbf{Z}(P) \leq H$. Consider $P/\mathbf{Z}(P)$. By induction $H/\mathbf{Z}(P)$ is properly contained in $\mathbf{N}(H/\mathbf{Z}(P))$. By the First Correspondence Theorem, then, $\mathbf{N}(H/\mathbf{Z}(P))$ maps back to $\mathbf{N}(H)$ in P and properly contains H. Hence the theorem holds for H.

Corollary 111a: Let P be a p-group. Then every maximal subgroup of P has index p and is normal in P.
Proof: is an exercise.

Exercise 118: Show that every subgroup and every quotient of a p-group is a p-group.
Exercise 119: Classify all the groups of order p^2 where p is a prime.
Exercise 120: Prove corollaries 109a and 109b.
Exercise 121: Prove corollaries 110a and 111a.

Lesson 30: Commutators

We can think of a commutator as taking one step beyond the conjugate to approximate commutativity. The commutator is generalized in each new level of algebra and always plays an important role.

Definition 64: Let G be any group and let g, h ∈ G be any elements. The **commutator** of g and h is the element computed by $g^{-1}h^{-1}gh = [g, h]$.

Note that $g^{-1}h^{-1}gh = (g^{-1}h^{-1}g)h = g^{-1}(h^{-1}gh)$, so a commutator is the product of a conjugate of g by the inverse of g, or the product of a conjugate of the inverse of h times h. [g, h] = 1 iff g and h commute with each other, so in an Abelian group all commutators are trivial. Equivalently [g, h] = 1 if and only if $\kappa_h(g) = g = g^h g$. We can also think of it this way: the commutator is a way of measuring how far conjugation removes an element from itself.

Note that commutation is a rule for combining two elements of G and producing a third element of G; that is, commutation is an operation on G. However, you can easily check that commutation is not an associative operation on a non-Abelian group: [[g, h], k] ≠ [g, [h, k]]. Also commutation is non-commutative: [g, h] ≠ [h, g] in a non-Abelian group. Commutators, in their more general definition, come into their own when we study non-associative algebras, most notably in Lie algebras. We will also define commutators of more than two arguments: [x, y, z] = [[x, y], z], and more generally we have $[x_1, x_2, x_3, \ldots, x_n] = [[x_1, x_2, x_3, \ldots, x_{n-1}], x_n]$. We use a shorthand notation [x, ny] = [x, y, y,..., y] where there are n y's in the second commutator.

We can pursue commutators by proving a seemingly endless series of commutator identities, but here we will focus on only the most useful and interesting ones. In every case the proofs are entirely straight-forward computations. Rather than list these identities as theorems we will refers to them as The Commutator Identities when we use them in the future.

> **The Commutator Identities**:
> 1. $[g,h]^{-1} = [h,g]$
>   It can be shown this way: $[g, h]^{-1} = (g^{-1}h^{-1}gh)^{-1} = h^{-1}g^{-1}hg = [h, g]$. We say that commutation is *anti-commutative*.
> 2. $[g, h^{-1}] = h[h, g]h^{-1}$
> 3. $[gh, k] = [g, k][g, k, h][h, k]$
> 4. $[g, hk][k, gh][h, kg] = 1$
> 5. $[kg, kh] = [g, hk][k, h]$
> 6. $(h^{-1}[g,h^{-1}, k] h)\cdot(k^{-1}[h, k^{-1},g] k)\cdot(g^{-1}[k, g^{-1},h] g) = 1$ This is called the **Hall-Witt Identity.**

There are a few identities that are worthy of singling out as theorems.

Theorem 112: $[g,k] = [h,k]$ iff $hg^{-1} \in Z_G(k^{-1})$.

Proof: [g,k] = [h,k] translates to $g^{-1}k^{-1}gk = h^{-1}k^{-1}hk$. Multiply both sides by h from the left, and then by k^{-1} on both sides from the right, and finally by g^{-1} on both sides from the right. Then give us $hg^{-1}k^{-1} = k^{-1}hg^{-1}$. Hence $hg^{-1} \in Z_G(k^{-1})$.

We can extend the definition of the commutator to subsets and subgroups in the natural way: let S, T ⊆ G and define [S, T] = {[s,t] | s ∈ S and t ∈ T}. *In general [S, T] is not a subgroup, because the product of two commutators is not necessarily a commutator.* The next theorem is the best we can do.

Theorem 113: If H ◁ G, and K ≤ G is any subgroup, then [H, K] ⊆ H; and if K is also normal in G, then [H, K] ⊆ H ∩ K.

Proof: [H, K] is the set of all elements of the form $h^{-1}k^{-1}hk$. Since $H \triangleleft G$, $k^{-1}hk = h' \in H$. Hence we have $[h, k] = h^{-1}k^{-1}hk = h^{-1}h' \in H$. Since h and k were arbitrary, $[H, K] \subseteq H$. The same argument shows that if K is normal then $[H, K] \subseteq K$ and therefore that $[H, K] \subseteq H \cap K$.

Since commutators are so closely related to conjugates, we should expect that commutators would be related to understanding conjugation maps, and therefore to be related to **Inn**(G).

Theorem 114: $\kappa_g \in \mathbf{Z}(\mathbf{Inn}(G))$ iff $[g^{-1}, G] \subseteq \mathbf{Z}(G)$.

Proof: First fix g. Then $\kappa_g \in \mathbf{Z}(\mathbf{Inn}(G))$ iff $\kappa_g \circ \kappa_h = \kappa_h \circ \kappa_g \ \forall \ \kappa_h \in \mathbf{Inn}(G)$
iff $\kappa_g(\kappa_h(x)) = \kappa_h(\kappa_g(x)) \ \forall x \in G$ and $\forall h \in G$
iff $g^{-1}h^{-1}xhg = h^{-1}g^{-1}xgh \ \forall x, h \in G$
iff $ghg^{-1}h^{-1}x = xghg^{-1}h^{-1} \ \forall x, h \in G$
iff $[g^{-1}, h^{-1}]x = x[g^{-1}, h^{-1}] \ \forall x, h \in G$
iff $[g^{-1}, h^{-1}] \in \mathbf{Z}(G) \ \forall h \in G$
iff $[g^{-1}, G] \subseteq \mathbf{Z}(G)$.

These commutator identities will prove useful in part 3. We will continue discussing commutators in the next lesson. We now have several results on the relationship between two subgroups. Some depend on both subgroups being normal, and some require that only one of the subgroups be normal. It is perhaps helpful to summarize what we know in a table.

If $H \leq G$ and $K \triangleleft G$	If $H \triangleleft G$ and $K \triangleleft G$	$\Longrightarrow$	and if also $H \cap K = 1$
then $H \cap K \triangleleft H$	then $H \cap K \triangleleft G$		then $hk = kh$
then $H \vee K = HK$	then $H \vee K \triangleleft G$		
then $(HK)/K \simeq H/(H \cap K)$		$\Longrightarrow$	and if further $HK = G$
then $[HK:K]$ and $[HK:K]$ are relatively prime			then $G \simeq H \times K$
then $[H,K] \subseteq K$	then $[H,K] \subseteq H \cap K$		
		$\Longrightarrow$	and if also K is minimal normal in G
			then $K \leq H$ or $H \vee K \simeq H \times K$
If $H \triangleleft S \leq G$ and $K \triangleleft G$ then $H \vee K \triangleleft S \vee K$			

Exercise 122: Check that commutation, as an operation on G, is neither associative, nor commutative if G is non-Abelian. Does it have an identity element? A unique one? Which elements have inverses?

Exercise 123: Prove the last five commutator identities listed above.

Exercise 124: Show that the following identities hold:
  a) $[g, hk] = [kg, h][gh, k]$
  b) $[hg, h^2] = [g, h^2]$
  c) $[g, [h, k]] = [g, k][k, hg][h, k]$
  d) $[g, [g, h]] = [g, h^{-1}gh]$
  e) $[g, h^2] = [hg, h][gh, h]$
  f) $[g, h^3] = [h^2g, h][hgh, h][gh^2, h]$. Now generalize (e) and (f) to a formula for $[g, h^n]$.
  g) derive an identity to re-express $[gk, hk]$ similar to identity 5.

Exercise 125: Show that $H \triangleleft G$ iff $[H, G] \leq H$.

Lesson 31: The Derived Subgroup

What is trivial in the Abelian case is decidedly non-trivial in the non-Abelian case. The product of two commutators may not equal another commutator. Hence, when we form the commutator of two subgroups, even if both of the subgroups are normal, the result is not always a subgroup. Because of this, many writers prefer to *define* the commutator [H, K] as *the subgroup generated by the set of all commutators* of elements of H with elements of K. There is a lot of convenience gained by setting up the definition like that, but I have elected not to do so. However, this means that when I wish to discuss the subgroup generated by a set of commutators I must either use the awkward notation < [H, K] > or introduce something new. At some point, the proliferation of new notation can become oppressive, but there is a matter of taste and convenience to consider when choosing notation. I use a different notation than most others.

Definition 65: Let G be any group. The subgroup generated by all the commutators of G is called the **derived subgroup** of G and is denoted by **dG** = < [G, G] >.

Usually the derived subgroup is denoted by G', but the prime notation is used for many other minor things and so I prefer to avoid it in this case. After the center of the group, the derived subgroup is the most useful subgroup. As the center of G is a measure of how commutative G is, the derived subgroup is a measure of how non-commutative the group is. The center and the derived subgroup are inverse to each other in meaning. For an Abelian group, G, we clearly have **dG** = **1** while **Z**(G) = G; Abelian groups are the least non-commutative and the most commutative, obviously. The larger the derived subgroup of G, the more entirely non-commutative it is so that when **dG** = G we have found a group which is as non-commutative as possible. On the other hand, the smaller the center of G, the less commutative it is so that when **Z**(G) = **1** we have found a group that is the least commutative a group can be. The two gauges of commutativity do not coincide. We may have a group that is the "least commutative" but not the "most non-commutative", and vice versa, though there are groups in which the two coincide. The center and the derived subgroups are measures to gauge commutativity only in a crude sense.

Theorem 115: For any group G, $dG \ll G$.

Proof: We need only show that every commutator is mapped to another commutator by any automorphism. So, let $\beta \in \mathbf{Aut}(G)$ be arbitrary and let $g, h \in G$ be any elements. Then
$$\beta([g, h]) = \beta(g^{-1}h^{-1}gh) = \beta(g^{-1})\beta(h^{-1})\beta(g)\beta(h) = \beta(g)^{-1}\beta(h)^{-1}\beta(g)\beta(h) = [\beta(g), \beta(h)]$$
So any automorphism maps commutators to other commutators. Since commutators are the generators of **dG**, we see that every automorphism of G preserves **dG** and therefore **dG** is a characteristic subgroup of G.

Corollary 115a: Let $H \triangleleft G$. Then $dH \triangleleft G$.
Proof: is an easy exercise.

We can use the presentation of a group to calculate its derived subgroup. For example, let's calculate $\mathbf{dD_n}$. Use the presentation $<r, d : r^n = d^2 = 1$, and $rd = dr^{-1} >$. There are four types of commutators that are possible:

$[r^k, r^m] = 1$ because the rotations all belong to the rotation subgroup $\mathbf{Z_n}$ which is Abelian.
$[r^k d, r^m] = (r^k d)^{-1} r^{-m} r^k d r^m = d r^{-k} r^{-m} r^k r^{-m} d = d r^{-2m} d = r^{2m}$
$[r^k d, r^m d] = (r^k d)^{-1} (r^m d)^{-1} r^k d r^m d = d r^{-k} d r^{-m} r^k d r^m d = d^2 r^k r^{-m} r^k r^{-m} d^2 = r^{2k-2m}$
$[r^k, r^m d] = (r^k)^{-1} (r^m d)^{-1} r^k r^m d = r^{-k} d r^{-m} r^k r^m d = d r^k r^{-m} r^k r^m d = r^{2k}$

Every commutator yields an even power of r, the rotational generator. When n is even, the even powers of r comprise all of $\mathbf{Z_{n/2}}$. When n is odd, the even powers of r comprise all of $\mathbf{Z_n}$. Thus we have

$\mathbf{dD_n} \simeq \mathbf{Z_n}$ if n is odd $\qquad\qquad \mathbf{dD_n} \simeq \mathbf{Z_{n/2}}$ if n is even

In particular $\mathbf{d}D_4 \simeq \mathbf{Z}_2$, $\mathbf{d}D_5 \simeq \mathbf{Z}_5$, and $\mathbf{d}D_6 \simeq \mathbf{Z}_3$. The property enjoyed by the derived subgroup that is the source of its importance to us is revealed in the following theorem.

Theorem 116: $\mathbf{d}G$ is the smallest subgroup of G for which $G/\mathbf{d}G$ is Abelian; more carefully, if $H \triangleleft G$ and G/H is Abelian, then $\mathbf{d}G \leq H$.

Proof: For convience, let $D = \mathbf{d}G$. Consider two elements of G/D, aD and bD. They commute if their commutator equals the identity in G/D. So we calculate:
$$[aD, bD] = (aD)^{-1}(bD)^{-1}(aD)(bD) = a^{-1}D\, b^{-1}D\, aD\, bD = a^{-1}b^{-1}abD = [a,b]D = D$$
because $[a, b] \in D$. Therefore aD commutes with bD and G/D is Abelian.

Now suppose $H \triangleleft G$ such that G/H is Abelian. Then for any $a, b \in G$ we know $aHbH = bHaH$. Then $abH = baH$. Hence $a^{-1}b^{-1}abH = H$, and therefore we have $a^{-1}b^{-1}ab = [a, b] \in H$. Since a and b were arbitrary, every commutator is included in H, and therefore $\mathbf{d}G \leq H$.

Essentially when we mod out $\mathbf{d}G$ we are modding out all of the non-Abelian-ness of G and what is left is the Abelian part of it. This is not how commutativity worked with the center. It was not the case that modding out the center modded out all of the Abelian-ness of G; $G/\mathbf{Z}G$ could very well be Abelian. $\mathbf{Z}(G)$ does not capture all of the Abelian-ness of G the way $\mathbf{d}G$ seems to capture all the non-Abelian-ness of it. In this sense, commutativity is somehow more fundamental to the nature of the group; we can't isolate all the commutativity of the group into a single subgroup the way we can isolate its non-commutativity.

By the Fundamental Theorem of Homomorphisms there exists a canonical homomorphism from G to $G/\mathbf{d}G$ whose kernal is $\mathbf{d}G$. This means that every group has a quotient which is Abelian. Though this Abelian quotient may be trivial, $\mathbf{d}G$ produces a quotient that is as large as possible. This leads to a new definition:

Definition 66: Let G be a group. $G/\mathbf{d}G$ is called the **abelianization** of G. The quotient is denoted by $G_{ab} = G/\mathbf{d}G$ and the canonical homomorphism is denoted by $\gamma_{ab} : G \to G/G'$.

You will be asked to show in the exercises that the derived subgroup of the alternating groups are themselves. In the case of the alternating groups, the measure of the center and the measure of the derived subgroup agree with each other. There is more terminology for this type of group.

Definition 67: When $G = \mathbf{d}G$ we say that G is a **perfect group.**

Theorem 117: Let G be any group and let $H, K \leq G$ be perfect subgroups. Then $H \vee K$ is a perfect subgroup.

Proof: $H \vee K$ is the subgroup generated by all products of elements of H with elements of K. Each element of H can be expressed as a product of commutators of elements of H because H is perfect. Similarly each element of K can be expressed as a product of commutators of elements of K. Hence we can express any element of $H \vee K$ as a product of elements of H and K and then substitute for each of those elements their expressions as commutators of elements in H or in K. Therefore $H \vee K$ is perfect.

Corollary 117a: Let G be any group. Then G has a maximal perfect subgroup which contains all other perfect subgroups of G.

Proof: is an exercise.

Definition 68: For any group G, the maximal perfect subgroup of G is called the **perfect radical** of G. This is not standard terminology. There is also no standard notation for the perfect radical of a group G, so I will adopt one: I will denote the perfect radical of G by $\mathbf{P}(G)$. This definition means that if H is any perfect subgroup of G, then $H \leq \mathbf{P}(G)$. In the exercises you will prove that $\mathbf{P}(G) \triangleleft G$. Naturally the perfect radical of a group may be trivial, as it always is for Abelian groups and a few others.

There is one more theorem to be included here for completeness, but its proof is left as an exercise.

Theorem 118: Assume $G \simeq H \times K$. Then $\mathbf{d}G \simeq \mathbf{d}H \times \mathbf{d}K$.

Proof: is an exercise.

Exercise 126: Show that $\mathbf{d}S_n \simeq A_n$, and that $\mathbf{d}A_n \simeq A_n$. Find $\mathbf{d}Q_{2^{\wedge}n}$.

Exercise 127: Prove corollary 115a.

Exercise 128: Prove corollary 117a.

Exercise 129: Prove theorem 118.

Exercise 130: Show that any subgroup conjugate to a perfect subgroup is perfect.

Exercise 131: Prove that $\mathbf{P}(G/\mathbf{P}(G)) = \mathbf{1}$.

Exercise 132: Show that $\mathbf{P}(G) \ll G$.

Exercise 133: Show that if K is a normal, cyclic subgroup of G then $\mathbf{d}G \leq Z_G(K)$.

Lesson 32: The Frattini Subgroup and non-Generators

We now come to the last of the four most important subgroups in a group. When we are trying to understand the structure of a large group, in general it is the bottom and the top parts of the lattice diagram that are the most accessible. The middle orders of the subgroups can be very numerous and intricately related, but it all narrows down to the trivial subgroup at the bottom and to the group itself at the top. By first looking at the cyclic subgroups generated by each of the elements, we determine the smallest order part of the structure; from those cyclic subgroups we can easily deduce any dihedral subgroups that may be present. We also looked at the minimal normal subgroups and have a considerable understanding of what some direct products and elementary subgroups that may be present. This gives us a good grasp of the lower part of the group. Then we turn our attention to the upper part of the group. We will now begin to consider what we can discover about the top part of a group, the largest of the subgroups. We will not complete the discussion of the largest subgroups until we get to the Sylow theorems in part 3. We will therefore look at the maximal subgroups, but this time we do not need to restrict ourselves to the normal ones.

Definition 69: Let G be any group and $H \leq G$. Suppose H has the property that whenever it is a subgroup of K then either $H = K$ or $K = G$. Then we say that H is a **maximal subgroup** of G.

Unlike the minimal subgroups, we have as yet no clear expectation about the nature of the maximal subgroups. At this point the value of considering the maximal subgroups comes in lowering our gaze down to their intersection.

Definition 70: The intersection of all the maximal subgroups of G is called the **Frattini subgroup** of G, and is denoted by $\Phi(G)$.

Some mathematicians denote the Frattini subgroup by **Frat**(G). These are easy to compute if we have a lattice diagram at our disposal, but otherwise they are challenging. Some are easy (refer back to the lattice diagrams to see these). Z_8 has only one maximal subgroup, Z_4, so $\Phi(Z_8) = Z_4$. In Z_{30} the maximal subgroups are Z_{15}, Z_{10}, and Z_6; hence $\Phi(Z_{30}) = 0$ (additive notation, remember). In $U_{15} \simeq Z_4 \times Z_2$ the maximal subgroups are the two copies of Z_4 and the copy of V. Thus we see that $\Phi(U_{15}) = Z_2$, the copy generated by 4. We can easily see, based on our previous work, that $\Phi(D_6) = <\rho_0>$ but that $\Phi(D_4) = <\rho_2>$. It is also easy to see that $\Phi(Q_8) = <a^2>$ and $\Phi(Q_{16}) = <x^2>$. Generally speaking, the Frattini subgroup will be pretty small, though it is constructed from the maximal subgroups. We use this lesson to prove two standard properties and an unexpected property of the Frattini subgroup.

Theorem 119: For any group G, $\Phi(G) \ll G$.

Proof: This proof is routine and is left as an exercise.

And as usual, we know the following:

Corollary 119a: If $H \triangleleft G$, then $\Phi(H) \triangleleft G$.

Proof: the same proof we have used for the center and the derived subgroup.

Before we get to the unexpected result, we need a definition.

Definition 71: Let G be any group. We will call an element $x \in G$ a **non-generator** of G if x has the property that whenever $<x, g_1, g_2, ..., g_n> = G$ then $<g_1, g_2, ..., g_n> = G$ equally well.

A non-generator of a group is an element that is never necessary to the construction of the group. It is important insofar as it exists, but it plays no role in determining the structure of the group. It is these non-generators that the Frattini subgroup captures for us.

Theorem 120: Let G be any group, then $\{ x \in G \mid x \text{ is a non-generator of } G \} = \Phi(G)$.

Proof: Let N be the set of non-generators of G. We will show that $\Phi(G) \subseteq N$ by contradiction. *Suppose there is some $g \in \Phi(G)$ and some subset $X \subseteq G$ such that $<X, g> = G$ but $<X> \neq G$; in other* words, suppose $\Phi(G)$ includes some element g that is not a non-generator. Clearly g is not an element of X. Since we are working with finite groups, it is clear that G has a subgroup M that is maximal with respect to the property that $X \subseteq M$ *but g is not in M*. This is another way of stating our supposition. At the very least M = < X >, but M could be larger. Now suppose $M \leq H \leq G$. Then $g \in H$ by the way we constructed M. But then $\{g\} \cup X \subseteq H$, and therefore $< g, X > \leq H$ and hence H = G by assumption. This shows that M is a maximal subgroup of G. But in that case $\Phi(G) \leq M$ and $g \in M$, contrary to our assumption. Therefore < X > = G and g is a non-generator for G. Therefore every element of $\Phi(G)$ is a non-generator and $\Phi(G) \subseteq N$.

We will now prove that $N \subseteq \Phi(G)$, again by contradiction. *Suppose $g \in N$ but that g is not an element of $\Phi(G)$*. Then g is not an element of some proper maximal subgroup M, and $M \neq < M, g >$. Since M is maximal this means that G = < M, g >. But since g is a non-generator for G, we must have G = M, contradicting our assumption that g is not an element of M. Thus $g \in M$ and hence is in every proper maximal subgroup of G. Therefore $g \in \Phi(G)$. Hence every non-generator is in $\Phi(G)$. We have shown that $N \subseteq \Phi(G)$ and therefore $\Phi(G) = N$.

The final property is not expected necessarily but is certainly welcome.
Theorem 121: $\Phi(H \times K) \simeq \Phi(H) \times \Phi(K)$.
Proof: This proof is left as an exercise, but as a hint use theorem 120.

In the next lesson, we will conclude this part of the modern algebra series by examining the role the Frattini subgroup plays in p-groups. These results will be useful in part 3.
Exercise 134: Find $\Phi(U_{21})$ and list the non-generators.
Exercise 135: Prove theorem 119.
Exercise 136: Prove theorem 121.

> Giovanni Frattini was born in 1852 and began studies at the University of Rome in 1869. He defined the subgroup named for him in a paper published in 1885. Though he pointed out that he had used a definition previously given by Alfredo Capelli the year before, and so it could fairly be called the Capelli subgroup, the current name has stuck. Frattini went on to devise an important method of proof used in group theory which is discussed in part 3 of this series and which was named the Frattini argument. Frattini died in Rome in 1925.
> Alfredo Capelli was born in 1855 in Milan, Italy, and went to study at the University of Rome in 1877. He worked as professor of mathematics in Pavia, Palermo, Bologna, and finally the University of Naples. He died in Naples in 1910.

Lesson 33: The Frattini Subgroup and p-groups

One class of groups in which Frattini subgroups play a significant role in are the p-groups. These groups were defined in definition 63 and will play a primary role in part 3. We know that all elementary Abelian groups are p-groups, but there are many p-groups that are not elementary. The only elementary Abelian 3-group of order 27 is $Z_3 \times Z_3 \times Z_3 = E_{27}$. $Z_9 \times Z_3$ is a 3-group of order 27 that is not elementary. We will consider in this lesson some of what the Frattini subgroup tells us about Abelian p-groups as the conclusion of this part of modern algebra and as something of a lead-in to part 3.

Theorem 122: Let P be any (finite) p-group, possibly non-Abelian, and let P' = P/Φ(P). Then P' is an elementary Abelian p-group. Further, if N is any normal subgroup such that P/N is elementary Abelian, then $\Phi(P) \leq N$.

Proof: It is immediate that P' is a p-group. We will show that every element in P' has order p. Let M be one of the maximal subgroups of P. Then [P:M] = p by corollary Corollary 111A. Now if $x \in M$ then clearly $x^p \in M$. If x is not an element of M, then M, xM, x^2M, …, x^{p-1}M is a complete collection of the cosets of M. Hence the pth power of every element of P is in M. Since M was an arbitrary maximal subgroup, x^p is in every maximal subgroup and therefore $x^p \in \Phi(P)$. Every element of P' has the form $x\Phi(P)$ and the identity element of P' is just $\Phi(P)$. Then for any element of P', $x\Phi(P)$, we can calculate $x^p\Phi(P) = (x\Phi(P))^p = \Phi(P)$. Thus every element of P' has order p. Therefore P' is a direct product of copies of Z_p and is an elementary Abelian p-group.

Corollary 122a: For any finite p-group G, $\mathbf{d}G \leq \Phi(P)$.
Proof: is left to the exercises.

In p-groups, then, all commutators are non-generators. Also the Frattini subgroup plays the same kind of role in quotients of p-groups that the derived subgroup plays in groups in general. **d**G is the smallest subgroup of G whose quotient G/**d**G is Abelian; and, for the class of p-groups at least, Φ(P) is the smallest subgroup of P whose quotient P/Φ(P) is elementary Abelian. It is a short step then to the next result:

Theorem 123: Let P be any (finite) p-group. If P' = P/Φ(P) is cyclic then P is cyclic.

Proof: If P' = P/Φ(P) is cyclic, then P' $\simeq Z_p$ since Z_p is the only cyclic elementary Abelian p-group. Thus we know $|\Phi(P)| = p^{n-1}$ since [P:Φ(P)] = p. Because Φ(P) has index p it is a maximal subgroup of P and therefore, by definition, it is the only maximal subgroup of P. Now the elements of Φ(P) are all non-generators of P. Hence P is generated by some elements not in Φ(P). Say x is one of the generators of P and suppose $< x > \neq P$. Then we must have $< x, \Phi(P) > \neq P$. Then $< x, \Phi(P) >$ is a proper subgroup of P containing Φ(P), contradicting that Φ(P) is a maximal subgroup of P. Therefore $< x > = P$ and P is cyclic.

Theorem 124: Let P be any (finite) p-group, possibly non-Abelian, and let $\beta \in$ **Aut**(P) have order q where q is a prime, $q \neq p$. If β fixes some coset of the Frattini subgroup, say $x\Phi(P)$, then β fixes at least one element of that coset.

Proof: Suppose $|P| = p^n$ and let $\beta \in$ **Aut**(P). Suppose β has order q where q is a prime and $q \neq p$. Finally suppose that β fixes some coset of Φ(P), say $x\Phi(P)$. We know that $|\Phi(P)| = p^k$ for some k < n (why can't k = n?). So $\beta(x\Phi(P)) = x\Phi(P)$. Let xf_0 be an arbitrary element of $x\Phi(P)$, so $\beta(xf_0) = xf_1$. Now if $xf_0 = xf_1$ then we have found an element fixed by β and are done. So we may suppose $xf_0 \neq xf_1$. Then $\beta^2(xf_0) = \beta(xf_1) = xf_2$, and again $xf_2 \neq xf_1$. Since the order of β is the prime q, $xf_2 \neq xf_0$ as well unless q = 2. We will come back to this possibility in a moment and proceed supposing $q \neq 2$. We may continue in this way getting images that are distinct from all the previous images until we finally arrive at $\beta^q(xf_0) = xf_0$, because q is the order of β. This gives a set of q distinct elements

$\{xf_0, xf_1, \ldots, xf_{q-1}\} = F_1 \subseteq x\Phi(P)$. This also shows that $q < p$.

We will now do the same thing again choosing $xf_0' \in x\Phi(P)$ but not in F_1. As above this will produce another set of q distinct elements $\{xf_0', xf_1', \ldots, xf_{q-1}'\} = F_2 \subseteq x\Phi(P)$. Since β is a bijection, F_1 and F_2 are disjoint. We can continue choosing elements of $x\Phi(P)$ which are not in any of the previous F_i and forming new distinct sets of q distinct elements; and we can do this m times where $mq < p^k < (m+1)q$. At no point will $mq = p^k$ since q and p are distinct primes, even when $q = 2$. Hence $p^k - mq > 0$, and this is the number of elements of $x\Phi(P)$ which we have not consider with respect to the action of β. But these remaining elements must be fixed by β for otherwise they would have to form a set of size q of consecutive images under β. Hence we have shown that β must fix at least one element of $x\Phi(P)$.

Now let's consider where we have come. We know that every group G possesses at least four characteristic subgroups: **soc**(G), **Z**(G), **d**G, and Φ(G). These four subgroups may each be trivial, and in general there is no reason to think they are closely connected with each other. But in the case of p-groups these four independently defined subgroups come in two pairs: we have **soc**(G) ≤ **Z**(G) and **d**G ≤ Φ(G). Whether there may be other connections between these subgroups we do not yet know.

There follows a table of all the groups up to order 32 that we have discussed in Part 1 or Part 2. You should be either familiar with each of them, having worked with them explicitly; or comfortable with them, recognizing them as similar to other groups you have worked with explicitly and remembering all the techniques necessary to a full understanding of them.

Exercise 137: Prove corollary 122a.

Table of Groups to Order 32 (Abelian groups in red and non-Abelian groups in black:)

order 2	Z_2					
order 3	Z_3					
order 4	$U_5 \simeq Z_4$	$U_8 \simeq V \simeq Z_2 \times Z_2 \simeq E_4$				
order 5	Z_5					
order 6	$U_7 \simeq U_9 \simeq Z_6$					
	$S_3 \simeq D_3$					
order 7	Z_7					
order 8	Z_8	$U_{24} \simeq Z_2 \times Z_2 \times Z_2 \simeq E_8$	$U_{15} \simeq U_{16} \simeq U_{20} \simeq U_{30} \simeq Z_4 \times Z_2$			
	D_4	Q_8				
order 9	Z_9	$Z_3 \times Z_3 \simeq E_9$				
order 10	$U_{11} \simeq Z_{10}$					
	D_5					
order 11	Z_{11}					
order 12 (1 missing)	$U_{13} \simeq Z_{12}$	$U_{21} \simeq U_{28} \simeq U_{36} \simeq U_{42} \simeq Z_6 \times Z_2$				
	$D_6 \simeq Z_2 \times D_3$	A_4				
order 13	Z_{13}					
order 14	Z_{14}					
	D_7					
order 15	Z_{15}					
order 16 (4 missing)	$U_{34} \simeq U_{17} \simeq Z_{16}$	$U_{32} \simeq Z_8 \times Z_2$	$Z_4 \times Z_4$			
	$U_{40} \simeq U_{48} \simeq U_{60} \simeq Z_4 \times Z_2 \times Z_2$		$Z_2 \times Z_2 \times Z_2 \times Z_2 \simeq E_{16}$			
	D_8,	Q_{16}	$Z_2 \times Q_8$	$Z_2 \times D_4$	M	
order 17	Z_{17}					
order 18 (1 missing)	$U_{19} \simeq U_{27} \simeq U_{54} \simeq Z_{18}$	$Z_6 \times Z_3$				
	D_9	$Z_3 \times D_3$				
order 19	Z_{19}					
order 20 (2 missing)	$U_{25} \simeq U_{50} \simeq Z_{20}$	$U_{33} \simeq U_{44} \simeq U_{66} \simeq Z_{10} \times Z_2$				
	$D_{10} \simeq Z_2 \times D_5$					
order 21 (1 missing)	Z_{21}					
order 22	$U_{23} \simeq U_{46} \simeq Z_{22}$					
	D_{11}					
order 23	Z_{23}					
order 24 (4 missing)	Z_{24}	$U_{39} \simeq U_{52} \simeq U_{78} \simeq U_{35} \simeq U_{45} \simeq U_{70} \simeq U_{90} \simeq Z_{12} \times Z_2 \simeq Z_6 \times Z_4$				
	$U_{56} \simeq U_{84} \simeq Z_6 \times Z_2 \times Z_2$					
	D_{12}	S_4	$Z_4 \times D_3$	$Z_3 \times D_4$	$Z_3 \times Q_8$	$Z_2 \times D_6$
	$Z_2 \times A_4$	$SL(2,3)$				
order 25	Z_{25}	$Z_5 \times Z_5$				
order 26	Z_{26}					
	D_{13}					
order 27 (2 missing)	Z_{27}	$Z_9 \times Z_3$	$Z_3 \times Z_3 \times Z_3 \simeq E_{27}$			
order 28 (1 missing)	$U_{29} \simeq U_{58} \simeq Z_{28}$	$Z_{14} \times Z_2$				
	$D_{14} \simeq Z_2 \times D_7$					
order 29	Z_{29}					
order 30	$U_{31} \simeq U_{62} \simeq Z_{30}$					
	D_{15}	$Z_3 \times D_5$	$Z_5 \times D_3$			
order 31	Z_{31}					
order 32 (35 missing)	Z_{32}	$U_{64} \simeq U_{51} \simeq Z_{16} \times Z_2$	$Z_8 \times Z_4$	$U_{96} \simeq U_{72} \simeq Z_8 \times Z_2 \times Z_2$		
	$U_{80} \simeq Z_4 \times Z_4 \times Z_2$	$Z_4 \times Z_2 \times Z_2 \times Z_2$	$Z_2 \times Z_2 \times Z_2 \times Z_2 \times Z_2 \simeq E_{32}$			
	D_{16}	Q_{32}	$Z_2 \times Q_{16}$	$Z_2 \times D_8$	$V \times Q_8$	$V \times D_4$
	$Z_4 \times D_4$	$Z_4 \times Q_8$	$Z_2 \times M$			

Index of Terms for Part 1 and Part 2

Abelian group	1.9, 2.5
Abelianization of a group	2.31
alternating groups	1.22, 1.23, 2.18, 2.27, 2.31
anti-commutative	2.30
anti-homomorphism	2.27
anti-isomorphism	2.21
associative law	1.2, 2.30
associative operation	1.2
automorphism	2.7, 2.8, 2.9, 2.10, 2.14
automorphism group	2.7, 2.8, 2.9, 2.10, 2.14, 2.33
bijection	1.4
bijective function	1.4
Butterfly Lemma	2.25
cancellation laws	1.6
canonical homomorphism	2.22, 2.31
Cauchy's Theorem	2.28
Cayley digraph	1.31, 2.4
Cayley table	1.11, 1.12
Cayley's Theorem	2.21
center	2.27, 2.28, 2.29, 2.30, 2.31
centerless group	2.27, 2.31
centralizer	2.28, 2.30
characteristic subgroup	2.10, 2.14, 2.19, 2.27, 2.31, 2.32
Chinese Remainder Theorem	2.6
commutative diagram	2.22
commutative law	1.9
commutative operation	1.9
commutator	2.30, 2.31
commuting elements	2.2, 2.13, 2.19
complex	2.12, 2.13
complex numbers	1.1
composition	1.2
congruence equation	1.11, 1.14, 1.18, 2.6
conjugacy class	2.26, 2.27, 2.28
conjugacy class equation	2.28, 2.29
conjugate element	2.14, 2.16, 2.26, 2.28, 2.30
conjugate subgroup	2.14, 2.15, 2.16, 2.26
conjugation map	2.14, 2.16, 2.30
Correspondence Theorems	2.23
coset	2.11, 2.12, 2.21
coset multiplication	2.20
coset representative	2.11
cube symmetries	1.27, 2.9
cycle	1.20
cycle notation	1.20

cycle type	2.16, 2.26
cyclic groups	1.9, 1.10, 1.14, 1.16, 1.17, 1.18, 2.1, 2.3, 2.5, 2.8, 2.27, 2.33
derived subgroup	2.31, 2.33
dihedral groups	1.24, 1.25, 1.26, 1.28, 2.5, 2.9, 2.13, 2.26, 2.27, 2.31
direct product	2.2, 2.3, 2.4, 2.5, 2.6, 2.10, 2.13, 2.19, 2.20, 2.22, 2.27, 2.31, 2.32
disjoint cycles	1.20
double coset	2.12, 2.15
elementary group	2.19, 2.29, 2.33
Euclidean algorithm	1.13
Euler's Phi function	1.15
even permutation	1.21, 1.22
exponential notation	2.14
extension (of a group)	2.17
factor group	2.2, 2.10
fiber	2.21
First Correspondence Theorem	2.23, 2.29
First Isomorphism Theorem	2.22, 2.24
fix (a letter)	1.19, 1.23
fraction	1.5
Frattini subgroup	2.32, 2.33
function diagram	1.8, 1.19, 1.21, 2.8
Fundamental Theorem of Homomorphism	2.22
generalized quaternions	1.30, 2.27, 2.29
generator	1.8, 1.9, 1.28, 2.8, 2.31
greatest common divisor	1.9, 1.11, 1.13
group	1.5
group of rigid motions	2.9
group of units	1.11, 1.12, 1.13, 1.14, 1.15, 1.16, 1.17, 1.18, 2.6, 2.8
Hall-Witt Identity	2.30
Hamiltonian groups	2.18
homomorphism	2.21, 2.22
identity element	1.3
imbedding	2.21
inclusion map	2.21
improper subgroup	1.7
index (of an element in U_n)	1.18
index (of a subgroup)	2.12, 2.23, 2.26, 2.28
injection	1.4
injective function	1.4
inner automorphism	2.14, 2.17, 2.20, 2.27, 2.27, 2.30
insertion	2.21
integers	1.1
internal direct product	2.5, 2.13
intersections	1.7, 1.9, 2.12, 2.15, 2.23, 2.24
inverse	1.4, 1.5, 1.13

isomorphism	2.1
join	1.7, 1.9, 2.12, 2.15, 2.19, 2.23, 2.24, 2.31
kernal	2.21, 2.22
Klein 4-group	1.11, 1.26, 1.28, 2.2, 2.19
Lagrange's Theorem	2.12
lattice diagram	1.10, 1.11, 1.12, 1.22, 1.23, 1.24, 1.25, 1.28, 2.1, 2.3, 2.4, 2.8, 2.9, 2.15, 2.17, 2.23, 2.24, 2.25
least common multiple	1.9
left coset	2.11
left identity	1.3
left inverse	1.4
left multiplication function	2.11, 2.21
length (of a cycle)	1.20
letter	1.19
maximal subgroups	2.29, 2.32
minimal normal subgroups	2.19, 2.29
modular law	1.10
Modular group	2.3
monoid	1.3, 1.4, 1.11
move (a letter)	1.19
n-cycle	1.20
natural homomorphism	2.22
natural numbers	1.1
non-generators	2.32
normal subgroup	2.13, 2.14, 2.15, 2.18, 2.20, 2.21, 2.22, 2.23, 2.24, 2.25, 2.26, 2.27, 2.31, 2.33
normalizer	2.26, 2.27, 2.28
odd permutation	1.21
operation on the set S	1.1, 2.30
order (of a group)	1.7, 1.9
order (of an element)	1.8, 1.9
outer automorphism	2.17, 2.20
p-group	2.6, 2.29, 2.33
partition (of a set)	2.11
partition (of an integer)	2.16
perfect group	2.31
perfect radical	2.31
permutation	1.19, 2.7, 2.16
permutation notation	1.19
power set	1.1, 1.2
presentation	1.28, 1.29, 1.30, 2.3, 2.17, 2.27
primitive root	1.16, 1.17
projection map	2.22
proper subgroup	1.7
quaternion group	1.29, 2.9, 2.18
quotient group	2.20, 2.22, 2.23, 2.24, 2.25, 2.27, 2.31, 2.33
rational numbers	1.1

real numbers	1.1
relation	1.28
right coset	2.11
right identity	1.3
right inverse	1.4
right multiplication function	2.11, 2.21
Second Correspondence Theorem	2.23
Second Isomorphism Theorem	2.24
self-normalizing subgroup	2.26
semi-group	1.2, 1.3
simple group	2.18, 2.19
socle	2.19, 2.29
special linear group	2.17
sporadic group	2.18
stabilizer	1.23, 2.7, 2.16
subgroup	1.7
subgroup generated by g	1.8, 1.9
subtraction	1.5
surjection	1.4
surjective function	1.4
symmetric group	1.19, 1.20, 1.23, 1.26, 2.13, 2.18, 2.21, 2.26, 2.27, 2.31
symmetry	1.24, 1.25, 2.9
Third Isomorphism Theorem	2.24
transposition	1.20, 1.21
trivial subgroup	1.7
two-sided identity	1.3
two-sided inverse	1.4
unit	1.11
Zassenhaus Theorem	2.25

Index of Symbols

A_n	1.22	$SL(n,m)$	2.17
$Aut(G)$	2.7	$soc(G)$	2.18
$\mathbb{C}$	1.1	$Stab(g)$	1.23
$\mathbb{C}^*$	1.1	U_n	1.11
$\mathbb{C}^*$	1.1	V	1.11
$C(g)$	2.26	$\times$	2.2
D_n	1.24	x^g	2.14
dG	2.31	$\mathbb{Z}$	1.1
$\mathbb{E}$	1.1	Z_n	1.9
E_n	2.19	$Z(G)$	2.27
$F(S)$	1.2	$Z_G(g)$	2.28
$Frat(G)$	2.32	γ_K	2.22
G'	2.29	γ_{ab}	2.31
G_{ab}	2.31	κ_g	2.14
$[G:H]$	2.12	λ_g	2.11
G/H	2.20	π_K	2.21
$<g>$	1.8	ρ_g	2.11
gH	2.11	$\varphi(n)$	1.15
Hg	2.11	$\phi\|_H$	2.10
$[H,K]$	2.30	$\Phi(G)$	2.32
$[h,k]$	2.30	$\in$	1.2
$\mathbb{I}$	1.1	$\vee$	1.7
$\mathbb{I}^0$	1.1	$\forall$	1.2
ind_r	1.18	$\simeq$	2.1
$Inn(G)$	2.14	$\hookrightarrow$	2.21
$ker(\phi)$	2.21	$\leq$	1.7
M	2.3	$\ll$	2.10
$mod(n)$	1.11	$\triangleleft$	2.13
$\mathbb{N}$	1.1		
$N(H)$	2.25		
$\mathbb{O}$	1.1		
O^*_{48}	1.27		
$Out(G)$	2.20		
$P(S)$	1.1		
$\mathbb{P}(G)$	2.31		
$\mathbb{Q}$	1.1		
$\mathbb{Q}^+$	1.1		
$\mathbb{Q}^*$	1.1		
Q_{2^n}	1.30		
$\mathbb{R}$	1.1		
$\mathbb{R}^+$	1.1		
$\mathbb{R}^*$	1.1		
S_n	1.20		

www.ingramcontent.com/pod-product-compliance
Lightning Source LLC
Chambersburg PA
CBHW051156220526
45473CB00003B/790